The Essentials of Data Science

Knowledge Discovery Using R

Chapman & Hall/CRC
The R Series

Series Editors

Aims and Scope

This book series reflects the recent rapid growth in the development and application of R, the programming language and software environment for statistical computing and graphics. R is now widely used in academic research, education, and industry. It is constantly growing, with new versions of the core software released regularly and more than 10,000 packages available. It is difficult for the documentation to keep pace with the expansion of the software, and this vital book series provides a forum for the publication of books covering many aspects of the development and application of R.

The scope of the series is wide, covering three main threads:

- Applications of R to specific disciplines such as biology, epidemiology, genetics, engineering, finance, and the social sciences.
- Using R for the study of topics of statistical methodology, such as linear and mixed modeling, time series, Bayesian methods, and missing data.
- The development of R, including programming, building packages, and graphics.

The books will appeal to programmers and developers of R software, as well as applied statisticians and data analysts in many fields. The books will feature detailed worked examples and R code fully integrated into the text, ensuring their usefulness to researchers, practitioners and students.

Published Titles

Stated Preference Methods Using R, *Hideo Aizaki, Tomoaki Nakatani, and Kazuo Sato*

Using R for Numerical Analysis in Science and Engineering, *Victor A. Bloomfield*

Event History Analysis with R, *Göran Broström*

Extending R, *John M. Chambers*

Computational Actuarial Science with R, *Arthur Charpentier*

Testing R Code, *Richard Cotton*

The R Primer, Second Edition, *Claus Thorn Ekstrøm*

Statistical Computing in C++ and R, *Randall L. Eubank and Ana Kupresanin*

Basics of Matrix Algebra for Statistics with R, *Nick Fieller*

Reproducible Research with R and RStudio, Second Edition, *Christopher Gandrud*

R and MATLAB® *David E. Hiebeler*

Statistics in Toxicology Using R *Ludwig A. Hothorn*

Nonparametric Statistical Methods Using R, *John Kloke and Joseph McKean*

Displaying Time Series, Spatial, and Space-Time Data with R, *Oscar Perpiñán Lamigueiro*

Programming Graphical User Interfaces with R, *Michael F. Lawrence and John Verzani*

Analyzing Sensory Data with R, *Sébastien Lê and Theirry Worch*

Parallel Computing for Data Science: With Examples in R, C++ and CUDA, *Norman Matloff*

Analyzing Baseball Data with R, *Max Marchi and Jim Albert*

Growth Curve Analysis and Visualization Using R, *Daniel Mirman*

R Graphics, Second Edition, *Paul Murrell*

Introductory Fisheries Analyses with R, *Derek H. Ogle*

Data Science in R: A Case Studies Approach to Computational Reasoning and Problem Solving, *Deborah Nolan and Duncan Temple Lang*

Multiple Factor Analysis by Example Using R, *Jérôme Pagès*

Customer and Business Analytics: Applied Data Mining for Business Decision Making Using R, *Daniel S. Putler and Robert E. Krider*

Flexible Regression and Smoothing: Using GAMLSS in R, *Mikis D. Stasinopoulos, Robert A. Rigby, Gillian Z. Heller, Vlasios Voudouris, and Fernanda De Bastiani*

Implementing Reproducible Research, *Victoria Stodden, Friedrich Leisch, and Roger D. Peng*

Graphical Data Analysis with R, *Antony Unwin*

Using R for Introductory Statistics, Second Edition, *John Verzani*

Advanced R, *Hadley Wickham*

The Essentials of Data Science: Knowledge Discovery Using R, *Graham J. Williams*

bookdown: Authoring Books and Technical Documents with R Markdown, *Yihui Xie*

Dynamic Documents with R and knitr, Second Edition, *Yihui Xie*

The Essentials of Data Science

Knowledge Discovery Using R

Graham J. Williams

CRC Press is an imprint of the
Taylor & Francis Group, an **informa** business
A CHAPMAN & HALL BOOK

CRC Press
Taylor & Francis Group
6000 Broken Sound Parkway NW, Suite 300
Boca Raton, FL 33487-2742

Printed in Canada on acid-free paper
Version Date: 20170616

International Standard Book Number-13: 978-1-138-08863-4 (Paperback)
International Standard Book Number-13: 978-1-4987-4000-5 (Hardback)

Visit the Taylor & Francis Web site at
http://www.taylorandfrancis.com

and the CRC Press Web site at
http://www.crcpress.com

For Catharina
Anam Cara

To Sean and Anita
Quiet lights that shine
Blessings

Preface

From data we derive information and by combining different bits of information we build knowledge. It is then with wisdom that we deploy knowledge into enterprises, governments, and society. Data is core to every organisation as we continue to digitally capture volumes and a variety of data at an unprecedented velocity. The demand for data science continues to growing substantially with a shortfall of data scientists worldwide.

Professional data scientists combine a good grounding in computer science and statistics with an ability to explore through the space of data to make sense of the world. Data science relies on their aptitude and art for observation, mathematics, and logical reasoning.

This book introduces the essentials of data analysis and machine learning as the foundations for data science. It uses the free and open source software R (R Core Team, 2017) which is freely available to anyone. All are permitted, and indeed encouraged, to read the source code to learn, understand, verify, and extend it. Being open source we also have the assurance that the software will always be available. R is supported by a worldwide network of some of the world's leading statisticians and professional data scientists.

Features

A key feature of this book, differentiating it from other textbooks on data science, is the focus on the hands-on end-to-end process. It covers data analysis including loading data into R, wrangling the data to improve its quality and utility, visualising the data to gain understanding and insight, and, importantly, using machine learning to discover knowledge from the data.

This book brings together the essentials of doing data science based on over 30 years of the practise and teaching of data science. It presents a programming-by-example approach that allows students to quickly achieve outcomes whilst building a skill set and knowledge base, without getting sidetracked into the details of programming.

The book systematically develops an end-to-end process flow for data science. It focuses on creating templates to support those activities. The templates serve as a starting point and can readily incorporate different datasets with minimal change to the scripts or programs. The templates are incrementally introduced in two chapters (Chapter 3 for data analysis and Chapter 7 for predictive machine learning) with supporting chapters demonstrating their usage.

Production and Typographical Conventions

This book has been typeset by the author using LaTeX and R's knitr (Xie, 2016). All R code segments included in the book are run at the time of typesetting the book and the results displayed are directly and automatically obtained from R itself.

Because all R code and screenshots are automatically generated, the output we see in the book should be reproducible by the reader. All code is run on a 64-bit deployment of R on a Ubuntu GNU/Linux system. Running the same code on other systems (particularly on 32 bit systems) may result in slight variations in the results of the numeric calculations performed by R.

Sample code used to illustrate the interactive sessions using R do not include the R prompt, which by default is "> ". Nor do they include the usual continuation prompt, which by default consists of "+ ". The continuation prompt is used by R when a single command extends over multiple lines to indicate that R is still waiting for input from the user. For our purposes, including the continuation prompt makes it more difficult to cut-and-paste from the examples in the electronic version of the book.

R code examples will appear as code blocks like the example code block shown over the page. The code block here uses `rattle::rattleInfo()` to report on the versions of the R software and many packages used at the time of compiling this book.

```
rattle::rattleInfo()

## Rattle: version 5.0.14 CRAN 4.1.0
## R: version 3.4.0 (2017-04-21)
##
## Sysname: Linux
## Release: 4.10.0-22-generic
## Version: #24-Ubuntu SMP Mon May 22 17:43:20 UTC 2017
## Nodename: leno
## Machine: x86_64
## Login: gjw
## User: gjw
## Effective_user: gjw
##
## Installed Dependencies
## ada: version 2.0-5
## amap: version 0.8-14
## arules: version 1.5-2
## biclust: version 1.2.0
## bitops: version 1.0-6
## cairoDevice: version 2.24
## cba: version 0.2-19
## cluster: version 2.0.6
## colorspace: version 1.3-2
## corrplot: version 0.77
## descr: version 1.1.3
## doBy: version 4.5-15
## dplyr: version 0.7.0
....
```

In providing example output from commands, at times long lines and long output will be replaced with ... and .... respectively. While most examples will illustrate the output exactly as it appears in R, there will be times where the format will be modified slightly to fit publication limitations. This might involve removing or adding blank lines.

The R code as well as the templates are available from the book's web site at `https://essentials.togaware.com`.

Currency

New versions of R are released regularly and as R is free and open source software a sensible approach is to upgrade whenever possible. This is common practise in the open source community, maintaining systems with the latest "patch level" of the software. This will ensure tracking of bug fixes, security patches, and new features.

The above code block identifies that version 3.4.0 of R is used throughout this book.

Acknowledgments

This book is a follow on from the Rattle book (Williams, 2011). Whilst the Rattle book introduces data mining with limited exposure to the underlying R code, this book begins the journey into coding with R. As with the Rattle book this book came about from a desire to share experiences in using and deploying data science tools and techniques through R. The material draws from the practise of data science as well as from material developed for teaching machine learning, data mining, and data science to undergraduate and graduate students and for professionals developing new skills.

Colleagues including budding and experienced data scientists have provided the motivation for the sharing of these accessible templates and reference material. Thank you.

With gratitude I thank my wife, Catharina, and children, Sean and Anita, who have supported and encouraged my enthusiasm for open source software and data science.

Graham J. Williams

Contents

List of Figures

List of Tables

1

Data Science

Over the past decades we have progressed toward today's capability to identify, collect, and store electronically a massive amount of data. Today we are *data* rich, *information* driven, and *knowledge* hungry, though, we may argue, *wisdom* scant. Data surrounds us everywhere we look. Data exhibits every facet of everything we know and do. We are today capturing and storing a subset of this data electronically, converting the data that surrounds us by *digitising* it to make it accessible for computation and analysis. We are digitising data at a rate we have never before been capable of. There is now so much data captured and even more yet to come that much of it remains to be analysed and fully utilised.

Data science is a broad tag capturing the endeavour of analysing *data* into *information* into *knowledge*. Data scientists apply an ever-changing and vast collection of techniques and technology from mathematics, statistics, machine learning and artificial intelligence to decompose complex problems into smaller tasks to deliver *insight* and *knowledge*. The knowledge captured from the smaller tasks can then be synthesised with *wisdom* to form an understanding of the whole and to drive the development of today's intelligent computer-based applications.

The role of a data scientist is to perform the transformations that make sense of the data in an evidence-based endeavour delivering the knowledge deployed with wisdom. Data scientists *resolve the obscure into the known.*[*] Such a synthesis delivers real benefit from the science—benefit for business, industry, government, environment, and humanity in general. Indeed, every organisation today is or should be a data-driven organisation.

---

[*] *Science is analytic description, philosophy is synthetic interpretation. Science wishes to resolve the whole into parts, the organism into organs, the obscure into the known.* (Durant, 1926)

A data scientist brings to a task a deep collection of computer skills using a variety of tools. They also bring particularly strong intuitions about how to tackle complex problems. Tasks are undertaken by resolving the whole into its parts. They explore, visualise, analyse, and model the data to then synthesise new understandings that come together to build our knowledge of the whole. With a desire and hunger for continually learning we find that data scientists are always on the lookout for opportunities to improve how things are done—how to do better what we did yesterday.

Finding such requisite technical skills and motivation in one person is rare—data scientists are truly scarce and the demand for their services continues to grow as we find ourselves every day with more data being captured from the world around us.

In this chapter we introduce the concept of the data scientist. We identify a progression of skill from the data technician, through data analyst and data miner, to data scientist. We also consider how we might deploy a data science capability.

With the goal of capturing knowledge as models of our world from data we consider the toolkits used by data scientists to do so. We introduce the most powerful software system for data science today, called R (R Core Team, 2017). R is open source and free software that is available to anyone and everyone. It offers us the freedom to use the software however we desire. Using this software we can discover, learn, explore, experience, extend, and share the algorithms for data science.

The Art of Data Science

As data scientists we ply *the art of excavating data for knowledge discovery* (Williams, 2011). As scientists we are also truly artists. Computer science courses over the past 30 years have shared the foundations of programming languages, software engineering, databases, artificial intelligence, machine learning, and now data mining and data science. A consistent theme has been that what we do as computer and data scientists is an art. Programming presents us with a language through which we express ourselves. We can use this language to communicate in a sophisticated manner with others. Our role is not to simply write code for computers to execute systematically but to express our views in an elegant form which we communicate both for execution by a computer and importantly for others to read, to marvel, and to enjoy.

As will become evident through the pages of this book, data scientists aim to not only turn data and information into intelligent applications, but also to gain insight and new knowledge from this data and information and to share these discoveries. Data scientists must also clearly communicate in such an elegant way so as to *resolve the obscure* and to make it *known* in a form that is accessible and a pleasure to read—in a form that makes us proud to share, want to read, and to continue to learn. This is the art of data science.

The Data Scientist

A data scientist combines a deep understanding of machine learning algorithms and statistics together with a strong foundation in software engineering and computer science and a well-developed ability to program with data. Data scientists cross over a variety of application domains and use their intuition to drive discoveries. As data scientists we experiment so as to deploy the right algorithm implemented within the right tool suite on the right data made available through the right infrastructure to deliver outcomes for the right problems.

The journey to becoming a data scientist begins with a solid background in mathematics, statistics and computer science and an enthusiasm for software engineering and programming computers. Their careers often begin as a data technician where skillful use of SQL and other data technologies including Hadoop are brought to bear to ingest and fuse data captured from multiple sources.

A data analyst adds further value to the extracted data and may rely on basic statistical and visual analytics supported by business intelligence software tools. A data analyst may also identify data quality issues and iterate with the data technician to explore the quality and veracity of the data. The role of a data analyst is to inform so as to support with evidence any decision making.

The journey then proceeds to the understanding of machine learning and advanced statistics where we begin to fathom the world based on the data we have captured and stored digitally. We begin to *program with data* in building models of the world that embody knowledge discoveries that can then improve our understanding of the world. Data miners apply a variety of tools to the increasingly larger volumes of data becoming more available in a variety of formats. By building models of the world—by learning from our interactions with the world captured through data—we can begin to understand and to build our knowledge base from which we can reason about the world.

The final destination is the art of data science. Data scientists are driven by intuition in exploring through data to discover the unknown. This is not something that can be easily taught. The data scientist brings to bear a philosophy to the knowledge they discover. They synthesise this knowledge in different ways to give them the wisdom to decide how to communicate and take action. A continual desire to challenge, grow and learn in the art, and to drive the future, not being pushed along by it, as the final ingredient.

It is difficult to be prescriptive about the intangible skills of a data scientist. Through this book we develop the foundational technical skills required to work with data. We explore the basic skill set for the data scientist.

Through hands-on experience we will come to realise that we need to program with our data as data scientists. Perhaps there will be a time when intelligent systems themselves can exhibit the required capabilities and sensitivities of today's most skilled data scientists, but it is not currently foreseeable. Our technology will continue to develop and we will be able to automate many tasks in support of the data scientist, but that intuition that distinguishes the skilled data scientist from the prescriptive practitioner will remain elusive.

To support the data scientists we also develop through this book two templates for data science. These scripts provide a starting point for the data processing and modelling phases of the data science task. They can be reused for each new project and will grow for each data scientist over time to capture their own style and focus.

Creating a Data Science Capability

Creating a data science capability can be cost-effective in terms of software and hardware requirements. The software for data science is readily available and regularly improving. It is also generally free (as in libre) and open source software (FLOSS). Today, even the hardware platforms need not be expensive as we migrate computation to the cloud where we can share resources and only consume resources when required.

The expense in creating a data science capability is in acquiring expert data scientists—bulding the team. Traditionally information technology organisations focused on delivering a centrally controlled platform hosted on premise by the *IT Department.* Large and expensive computers running singularly vetted and extremely expensive statistical software suites were deployed. Pre-specified requirements were provided through a tender process which often took many months or even years. The traditional funding models for many organisations preferred on-premise expenses instead of otherwise much more cost-effective, flexible and dynamic data science platforms combing FLOSS with cloud.

The key message from many years of an evolving data science capability is that the focus must be on the skills of the practitioner more than the single vendor provided software/hardware platform. Oddly enough this is quite obvious yet it is quite a challenge for the era mentality of the IT Department and its role as the director rather than the supporter of business. Recent years have seen the message continue to be lost. Slowly though we continue to realise the importance business driving data science rather than IT technology being the driver.

The principles of the business drivers allowing the data scientists to direct the underlying support from IT, rather than vice-versa, were captured by the Analyst First* movement in the early 2000s. Whilst we still see the technology first approach driven by

*`http://analystfirst.com/core-principles/`

vested interests many organisations are now coming to realise the importance of placing business driven data science before IT.

The Analyst First movement collected together principles for guiding the implementation of a data science capability. Some of the key principles, relevant to our environment today, can be paraphrased as:

- A data science team can be created with minimal expense;
- Data science, done properly, is scalable;
- The human is the most essential, valuable and rare resource;
- The scientist is the focus of successful data science investment;
- Data science is not information technology;
- Data scientists require advanced/flexible software/hardware;
- There is no "standard operating environment" for data science;
- Data science infrastructure is agile, dynamic, and scalable.

It is perhaps not surprising that large organisations have struggled with deploying data science. Traditional IT departments have driven the provision of infrastructure for an organisation and can become disengaged from the actual business drivers. This has been their role traditionally, to source software and hardware for specific tasks as they understand it, to go out to tender for the products, then provision and support the platform over many years.

The traditional approach to creating a data science team is then for the IT department, driven by business demands, to educate themselves about the technology. Often the IT department will invite vendors with their own tool suites to tender for a single solution. A solution is chosen, often without consulting the actual data scientists, and implemented within the organisation. Over many years this approach has regularly failed.

It is interesting to instead consider how an open source product like R[*] has become the tool of choice today for the data scientist. The open source community has over 30 years of experience in delivering powerful excellent solutions by bringing together skilled and passionate developers with the right tools. The focus is on allowing these solutions to work together to solve actual problems.

Since the early 1990s when R became available its popularity has grown from a handful of users to perhaps several million users today. No vendor has been out there selling the product. Indeed, the entrenched vendors have had to work very hard to retain their market position in the face of a community of users realising the power of open source data science suites. In the end they cannot compete with an open source community of many thousands of developers and statisticians providing state-of-the-art technology through free and open source software like R. Today data scientists themselves are driving the technology requirements with a focus on solving their own problems.

The world has moved on. We need to recognise that data science requires flexibility and agility within an ever-changing landscape. Organisations have unnecessarily invested millions in on-premise infrastructure including software and hardware. Now the software is generally available to all and the hardware can be sourced as and only when required.

Within this context then open source software running on computer servers located in the cloud delivers a flexible platform of choice for data science practitioners. Platforms in the cloud today provide a completely managed, regularly maintained and updated, secure and comprehensive environment for the data scientist. We no longer require significant investment in corporately managed, dedicated and centrally controlled IT infrastructure.

Closed and Open Source Software

Irrespective of whether software can be obtained freely through a free download or for a fee from a vendor, an important requirement for innovation and benefit is that the software source codes

[*] R, a statistical software package, is the software we use throughout this book.

be available. We should have the freedom to review the source code to ensure the software implements the functions correctly and accurately and to simply explore, learn, discover, and share. Where we have the capability we should be able to change and enhance the software to suit our ever-changing and increasingly challenging needs. Indeed, we can then share our enhancements with the community so that we can build on the shoulders of what has gone before. This is what we refer to by the use of *free* in free (as in libre) open source software (FLOSS). It is not a reference to the cost of the software and indeed a vendor is quite at liberty to charge for the software.

Today's Internet is built on free and open source software. Many web servers run the free and open source Apache software. Nearly every modem and router is running the open source GNU/Linux operating system. There are more installations of the free and open source Linux kernel running on devices today than any other operating system ever—Android is a free and open source operating system running the Linux kernel. For big data Hadoop, Spark, and their family of related products are all free and open source software. The free and open source model has matured significantly over the past 30 years to deliver a well-oiled machine that today serves the software world admirably. This is highlighted by the adoption of free and open source practises and philosophies by today's major internet companies.

Traditionally commercial software is closed source. This presents challenges to the effective use and reuse of that software. Instead of being able to build on the shoulders of those who have gone before us, we must reinvent the wheel. Often the wheel is re-implemented a multitude of times. Competition is not a bad thing per se but closed source software generally hinders progress. Over the past two decades we have witnessed a variety of excellent machine learning software products disappear. The efforts that went into that software were lost. Instead we might recognise business models that compensate for the investments but share the benefits and ensure we can build on rather then reinvent.

The Data Scientist's Toolkit

Since the development of the free and open source R statistical software in 1995 it has emerged to be the most powerful software tool for programming over data. Today it provides all of the capabilities required for turning data into information and then information into knowledge and then deploying that knowledge to drive intelligent applications.

R is a key tool within the modern data scientist's toolkit. We will often combine the data processing and statistical tools of R with the powerful command line processing capabilities of the Linux ecosystem, for example, or with other powerful general purpose programming languages such as Python and specialist languages like SQL. The importance of R cannot be understated.

The complementary nature of the open source statistical language and the open source operating system combine to make R on Linux and particularly Ubuntu a most powerful platform for data science.* We only need to note that cloud offerings of pre-configured linux-based data science virtual machines are now common and provide within 5 minutes a new server running a complete free and open source software stack for the data scientist. The virtual machines running on the cloud can be any of a variety of sizes and can be resized as required and powered down when not required and so incurring minimal cost.

We will focus on R as the programming language for the data scientist, programming with data, using software engineering skills to weave narratives from the data through analysis. We will proceed on to the world of R to support hands-on data science and a process for delivering data science

Future: Massive Ensembles and Extreme Distribution

We finish our introduction with a glimpse of a future world of data. We envisage massive ensembles of distributed models communicating to support intelligent applications working with extremely distributed data that exists across a multitude of sources (Zhou *et al.*, 2014).

---

*R can also be effectively deployed on Max/OSX and Windows.

For over 20 years Internet companies have separately collected massive stores of personal and private data from users of their software and devices. We have been quite happy to provide our data to these organisations because of the perceived and actual benefits from doing so. Data today is apparently owned or at least managed by Google, Apple, Microsoft, Facebook, Amazon and many others. Through the services they provide we have been able to discover old friends, to share our ideas and daily activities publicly, to have our email filtered of spam and managed for us and available from a variety of devices. To have our diaries and calendars actively supporting us and purchase books and music.

With widespread availability of massive volumes of data coupled with powerful compute today we can build intelligent applications that learn as they interact with the world. Google and Microsoft have been leading the way in making use of this data through the application of artificial intelligence technologies, and in particular machine learning algorithms and through the recent emergence of deep learning algorithms. The latter technology matches massive data with massive computational resources to learn new types of models that capture impressive capability in specific areas.

There are two parallel problems that we need to address though: privacy and distribution. Privacy is a serious concern as many have argued over many years. Whilst we may be willing to share our data with today's organisations and governments, can we have an ongoing trust of the data custodians and of the principles of these organisations that may well change over time?

Even now our private data with our agreement is made available to many organisations, purposefully by sharing it for commerce and/or inappropriately through security breeches. The concept of a centralised collection of personal and private data is regularly challenged.

We are now seeing a growing interest and opportunity to turn this centralized mode of data collection on its head, and to return the focus to privacy first.

Already many governments have strong laws that prohibit or limit the sharing of data across agencies and between organisations within their jurisdiction. These organisations each have evolving snapshots of data about an ever-increasing overlap of individuals. An individual usually has the right to access this data across these different organisations. They may be able to access their own bank data, medical records, government collected tax data, personal holiday data, history of shopping online, location tracking data for the past 5 years, personal collection of music and movies, store of photos, and so much more. All of this data is spread across a multitude of organisations. Those organisations themselves generally cannot or do not share the data.

Envisage a world then that begins to see a migration of data from central stores to personal stores. As we migrate or augment on-premise stored data to cloud stored data, we will also see applications that retain much personal data within our own personal stores but collected on the cloud with unbreakable encryption. This will result in an extreme data distribution and presents one of the greatest research challenges for data science in the near future.

Envisage being able to access all of this data relating to yourself and bring this data together so that increasingly intelligent personal applications can know and respond in a very personal manner. The complete data store and applications live on their own devices under the owner's control. The intelligent applications will interact with a myriad of other personal and corporate and government intelligent applications across the cloud, retaining privacy supported by intelligent applications from a variety of organisations adding their own value but in a personal context on personal devices.

Finale

This is a future scenario to ponder and to create. A world with eight billion or more individually evolved intelligent applications. Applications that interact across an incredibly complex but intelligent and interconnected cloud.

Through this book we simply provide the beginnings of a journey towards data science and knowledge discovery.

1.1 Exercises

Exercise 1.1 What is Data Science?

We have presented one view of data science but there are many views about what data science is. Explore other definitions of data science and compare and contrast data science with data mining, statistics, analytics, and machine learning. Write a short position brief that summarises the various standpoints and leads to your view of what data science is about.

Exercise 1.2 Who is a Data Scientist?

A skill set for a data scientist is proposed and as with the term data science there are many views. Explore other views of the skill sets that are required for a data scientist. Write a short position brief that summarises the various standpoints. Include your view of what a data scientist is.

Exercise 1.3 Data Scientist's Toolkit

Open source software is extensively deployed in data science. Research the tools that are available, both open and closed source. Identify and comment upon the pros and cons reported for the different tools in a short position brief.

Exercise 1.4 Open Source Data

The importance of open source is argued in this chapter. The argument relates specifically to open source software but is equally important to open source data. Investigate the importance of open source data and identify where open source data can readily be found across the Internet.

Exercise 1.5 Open Source Policy

The concept of open source policy aims to provide transparency in the assumptions and the modelling conducted to support different government policy agendas. Discuss the practicality and the benefits of open source models for policy. References might include Henrion (2007) and Lobo-Pulo (2016).

2

Introducing R

Data scientists spend a considerable amount of time writing programs to ingest, fuse, clean, wrangle, visualise, analyse, model, evaluate, and deploy models from data. We rely on having the right tools to allow us to interact with data at a sophisticated level which usually involves a language through which we express our narrative. Throughout this book the concept of formally writing sentences in such a language is introduced. Through these formal sentences we tell the story of our data.

Like other classes of language a programming language allows us to express and communicate ourselves. Programming languages tend to be more strictly specified than human languages. To some extent all languages follow rules, referred to as the syntax (which tells us the rules for constructing sentences in the language) and semantics (which tells us what the words of the language mean). Unlike human languages, programming languages require all of their carefully crafted rules to be precisely followed.

Through this book we will learn a language for programming with data. There are many programming languages we can choose from including Python which is well established by computer scientists as a carefully crafted instrument for expressing ourselves. A less well-crafted programming language and one that has grown organically through its power to express ideas about data succinctly has emerged from the statistical community. This language has borrowed concepts originally from the artificial intelligence community and significantly from the computer science community over the years. That language is R (R Core Team, 2017).

Our journey to understanding data progresses by capturing a narrative around that data through this language. We will write sentences that delve into the data and deliver insights and understanding. Our narrative expressed in and resulting from our programming is built from sentences, paragraphs, and complete stories.

To begin this journey to data science through R we assume basic familiarity and experience with the syntax and semantics of R. This level of understanding can be gained through hands on experience whilst working with the many resources available on the Intranet. Particularly recommended is to work through the standard *Introduction to R* available from `https://cran.r-project.org/manuals.html`. A gentle introduction to programming with R for data mining is also available through the graphically based R code generator Rattle (Williams, 2009, 2011).

Our journey will introduce the syntax and semantics of the language as we proceed. The introduction to R in this book is not extensive and so it is recommended that when you come across some syntax or function that is not explained be sure to seek understanding before continuing.

An important motivation as we proceed through this book is to develop useful sentences in our language as we go. We call sentences from a programming language *code*. The codes from the language express what we ask a computer to do. However, they also capture what we wish to communicate to other readers. A collection of sentences written in such a language will constitute what we call a *program*.

Examples will be used extensively to illustrate our approach to programming over data. Programming by example is a great way to continually learn, by watching and reviewing how others speak in that language and to pick up the idioms and culture of a language. Our aim is to immerse ourselves in the process of writing programs to deliver insights and outcomes from our data. We encourage *programming by example* so as to build on the shoulders of those who have gone before us.

We will learn to use a tool that supports us in our programming. We will also introduce a process to follow in exploring our own data. The aim is that through *learning by immersion* and *programming by example* we gain an increasing understanding of the syntax and semantics of the language of data science. And as we proceed to deliver new insights into our data we will develop a process that we will find ourselves repeating with new tasks.

The chapters that follow this introduction will fill in much of the detail and more of the understanding of the language. They will lead us to understand some of the process of data science and of the verbs (the action words or commands) that we use in data science to accomplish our goals—to deliver the outcomes that we have set for ourselves.

We recognise that R is a large and complex ecosystem for the practice of data science. It is an evolving language, evolving as we learn to achieve more with it in better ways. As practitioners of the language we will often find ourselves evolving with it, as new capabilities and simpler grammars become available. We have certainly seen this with the more recent availability of new grammar-focused developments of the language.*

As data scientists we will continue to rely on the increasing wealth of information available as we grow our experiences. There is a breadth of freely available information on the Internet from which we can continually learn. Searching the Internet will lead us to useful code segments that illustrate almost any task we might think of. Exploring such information as we proceed through the material in this book is strongly encouraged.

Finally, this book provides a quick start guide to working with data as data scientists. Once you have established the foundations introduced here there are many more detailed journeys you can take to discover the full power of the language you have begun to learn. You will find yourself travelling through the literature of R to discover delightful new ways of expressing yourself. This is the journey to data science. Through data science we will find a foundation from artificial intelligence and machine learning that will lead on to intelligent applications.

In this introductory chapter on R we cover some of the basics. This includes identifying a platform through which we interact with R called RStudio. We will introduce the concepts of libraries of packages which considerably extend the language itself to deliver an extensive catalogue of capability. An overview of the con-

---

*https://github.com/tidyverse

cepts of functions, commands and operators will provide the fundamentals for understanding the capabilities of R. We conclude the chapter by introducing the powerful concept of pipes as a foundation for building sophisticated data processing pipelines using a series of simple operations to achieve something quite complex. We also point to the extensive collections of R documentation that is available including both formal documentation and the extensive crowd sources resources (such as stack overflow).

With these basics in hand the remaining chapters introduce R programming concepts oriented toward the common tasks of a data scientist. Whilst these can be followed through sequentially as they are, a recommended approach is to actually jump to Chapter 10 to learn about a discipline or a process for doing literate data science. Through this process we capture our narrative—the story that the data tells—intertwined with the data that actually supports the narrative. I would also encourage you to review Chapter 11 where a style for programming in R is presented. You are encouraged to strictly follow that style as you begin to develop your own programs and style.

2.1 Tooling For R Programming

The software application known as the R interpreter is what interprets the programs that we write in R. It will need to be installed on our computer. Usually we will install R on our own computer and instructions for doing so are readily available on the Internet. The R Project* is a good place to start. Most GNU/Linux distributions provide R packages from their application repositories. Now is a good time to install R on your own computer.

The free and open source RStudio† software is recommended as a modern **integrated development environment** (IDE) for writing R programs. It can be installed on all of the common desktop operating systems (Linux, OSX, Windows). It can also

*https://www.r-project.org
†https://www.rstudio.com

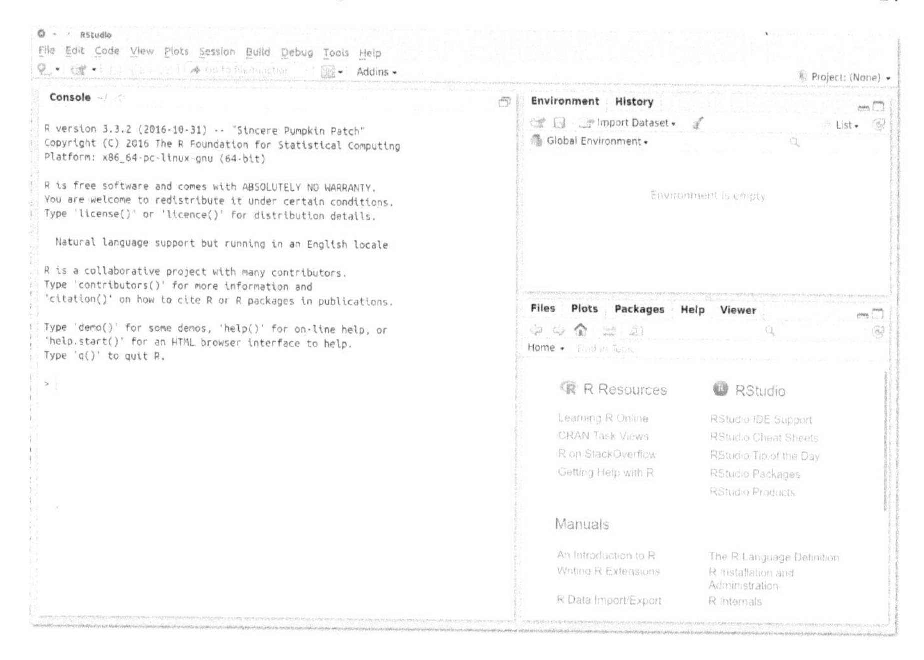

Figure 2.1: *The initial layout of the* RStudio *window panes showing the* `Console` *pane on the left with the* `Environment` *and* `Help` *panes on the right.*

be installed on a server running the GNU/Linux operating system which then supports browser-based access to R running on a back-end cloud server.

Figure 2.1 shows the RStudio interface as it might appear for the first time after it has been installed. The interface includes an R `Console` on the left through which we directly communicate with the R interpreter. The window pane on the top right provides access to the `Environment` which includes the data and datasets that are defined as we interact with R. Initially it is empty. A `History` of all the commands we have asked R to run is available on another tab within the top right pane.

The bottom right pane provides direct access to an extensive collection of `Help`. On another tab within this same pane we can access the `Files` on our local storage. Other tabs provide access to `Plots`, `Packages`, and a `Viewer` to access documents that we might generate from the application.

In the Help tab displayed in the bottom right pane of Figure 2.1 we can see numerous links to R documentation. The Manual titled *An Introduction to R* within the Help tab is a good place to start if you are unfamiliar with R. There are also links to learning more about RStudio and these are recommended if you are new to this environment.

The server version of RStudio runs on a remote computer (e.g., a server in the cloud) with the graphical interface presented within a web browser on our own desktop. The interface is much the same as the desktop version but all of the commands are run on the server rather than on our own desktop. Such a setup is useful when we require a powerful server on which to analyse very large datasets. We can then control the analyses from our own desktop with the results sent from the server back to our desktop whilst all the computation is performed on the server. Typically the server is a considerably more powerful computer than our own personal computers.

Often we will be interacting with R by writing code and sending that code to the R Interpreter so that it can be run (locally or remotely). It is always good practice to store this code into a file so that we have a record of what we have done and are able to replicate the work at a later time. Such a file is called an R Script. We can create a new R Script by clicking on the relevant icon on the RStudio toolbar and choosing the appropriate item in the resulting menu as in Figure 2.2. A keyboard shortcut is also available to do this: `Ctrl+Shift+N` (hold the `Ctrl` and the `Shift` keys down and press the `N` key). A new file editor is presented as the top left pane within RStudio. The tab will initially be named `Untitled1` until we actually save the script to a file. When we do so we will be asked to provide a suitable name for the file.

The editor is where we write our R code and compose the programs that instruct the computer to perform particular tasks. The editor provides numerous features that are expected in a modern program editor. These include syntax colouring, automatic indentation to improve layout, automatic command completion, interactive command documentation, and the ability to send specific commands to the R Console to have them run by the R Interpreter.

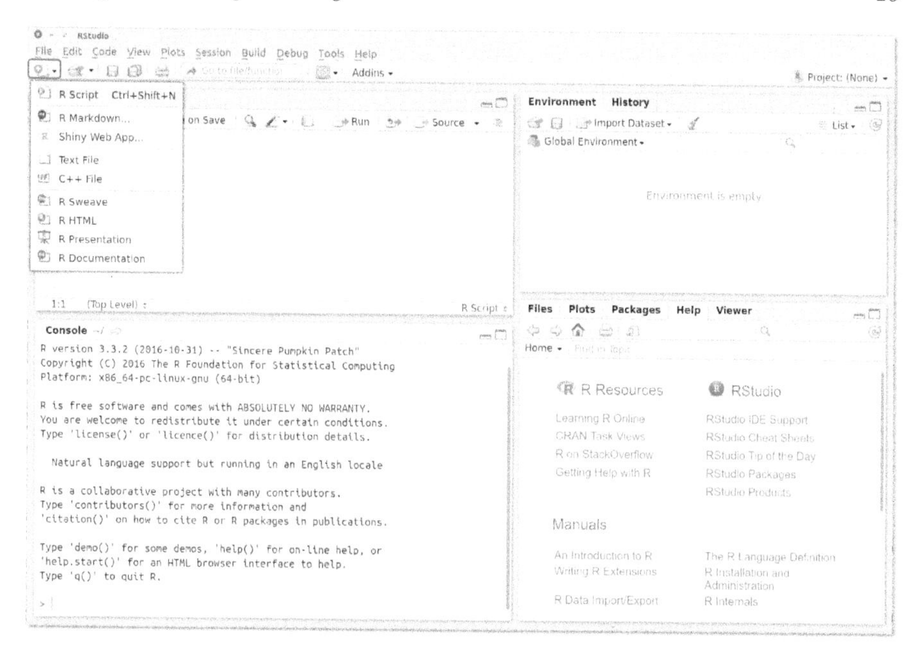

Figure 2.2: *Ready to edit R scripts in RStudio using the* **`Editor`** *pane on the top left created when we choose to add a new R Script from the toolbar menu displayed by clicking the icon highlighted in red.*

We can ask RStudio to send R commands to the R Console through the use of the appropriate toolbar buttons. One line or a highlighted region can be sent using the **`Run`** button found on the RStudio toolbar as highlighted in Figure 2.3. Having opened a new R script file we can enter commands like those below. The example shows four commands that together are a program which instructs the R interpreter. The first command is `install.packages()` which ensures we have installed two requisite software packages for R. We only need to do this once and the packages are available from then on when we use R. The second and third commands use the `library()` command to make these software packages available to the current R session. The fourth command produces a plot using `qplot()` using data from the `weatherAUS` dataset provided by rattle (Williams, 2017). This dataset captures observations of

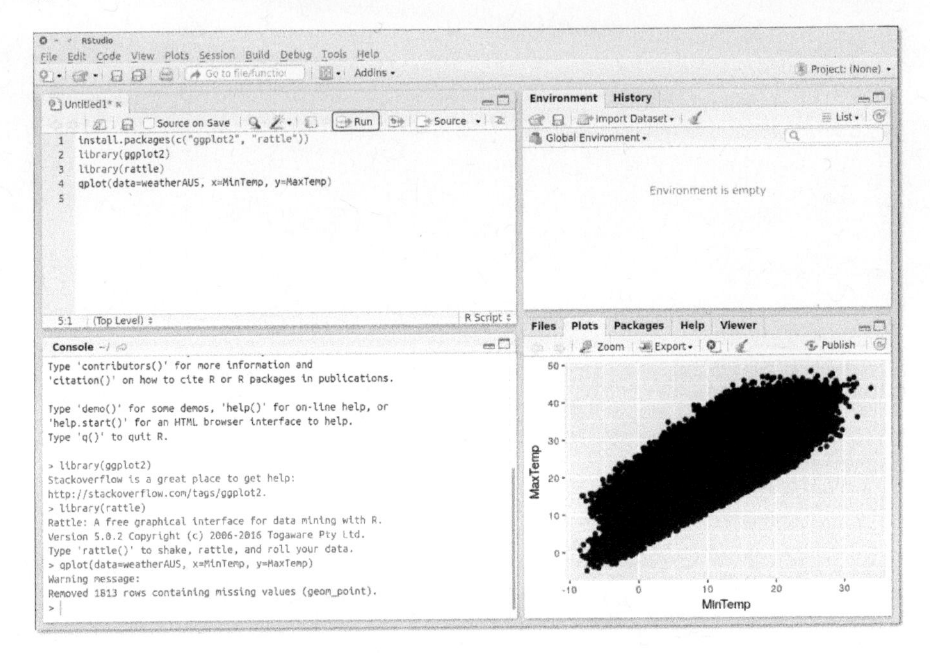

Figure 2.3: *Running* R *commands in* RStudio*. The* R *programming code is written into a file using the editor in the top left pane. With the cursor on the line containing the code we click the* **Run** *button to pass the code on to the* R **Console** *to have it run by the* R *Interpreter to produce the plot we see in the bottom right pane.*

weather-related variables for over 8 years across almost 50 weather stations in Australia.

```
install.packages(c("ggplot2", "rattle"))
library(ggplot2)
library(rattle)
qplot(data=weatherAUS, x=MinTemp, y=MaxTemp)
```

The `qplot()` command from the ggplot2 (Wickham and Chang, 2016) package allows us to quickly construct a plot—hence its name. From the `weatherAUS` dataset we choose to plot the minimum daily temperature (`MinTemp`) on the x-axis against the maximum daily temperature (`MaxTemp`) on the y-axis. The resulting plot is called a **scatter plot** and we see the plot in the lower right pane of Figure 2.3. It's a rather black blob for now but as we proceed we will learn a variety of techniques for visualising the data effectively. Already though we gain insight about the relationship between the daily minimum and maximum temperatures. There appears to be a strong linear relationship.

Figure 2.3 also shows the RStudio editor as it appears after we type the above commands into the R `Script` file in the top left pane. We have sent the commands to the R `Console` to have it run by R. We have done this by ensuring the cursor within the R `Script` editor is on the same line as the command to be run and then clicking the **`Run`** button. We will notice that the command is sent to the R `Console` in the bottom left pane and the cursor advances to the next line within the R `Script`. After each command is run any text output by the command is displayed in the R `Console` which might simply be informative messages about a package or warnings and errors that arise in processing the data. Graphic output is displayed in the `Plots` tab of the bottom right pane.*

We have now written our first program in R and can provide our first observations of the data. It is not too hard to see from the plot that there appears to be quite a strong relationship between the minimum temperature and the maximum temperature: with higher values of the minimum temperature recorded on any particular day we see higher values of the maximum temperature. There is also a clear lower boundary that might suggest, as logic would dictate, that the maximum temperature cannot be less than the minimum temperature. If we were to observe data points below this line then we would begin to explore issues with the quality of the data.

As data scientists we have begun our observation and understanding of the data, taking our first steps toward immersing ourselves in and thereby beginning to understand the data.

---

*It is opportune for the reader to replicate this program for themselves using their own installation of R and RStudio.

2.2 Packages and Libraries

The power of the R ecosystem comes from the ability of the community of users to themselves extend the language by its nature as **open source software**. Anyone is able to contribute to the R ecosystem and by following stringent guidelines they can have their contributions included in the comprehensive R archive network, known as CRAN.* Such contributions are collected into what is called a *package*. Over the decades many researchers and developers have contributed thousands of *package*s to CRAN with over 10,000 packages available for R from almost as many different authors.

A ***package*** is how R collects together *command*s for a particular task. A ***command*** is a verb in the computer language used to tell the computer to do something. Hence there are very many verbs available to build our sentences to command R appropriately. With so many packages there is bound to be a package or two covering essentially any kind of processing we could imagine. We will also find packages offering the same or similar commands (verbs) perhaps even with very different meanings.

Beginning with Chapter 3 we will list at the beginning of each chapter the R *package*s that are required for us to be able to replicate the examples presented in that chapter. Packages are installed from the Internet (from the securely managed CRAN package repository) into a local *library* on our own computer. A ***library*** is a folder on our computer's storage which contains sub-folders corresponding to each of the installed packages.

To install a package from the Internet we can use the command `install.packages()` and provide to it as an *argument* the name of the package to install. The package name is provided as a ***string*** of ***characters*** within quotes and supplied as the `pkgs=` ***argument*** to the command. In the following we choose to install a package called dplyr (Wickham *et al.*, 2017a)—a very useful package for data wrangling.

---

*`https://cran.r-project.org`.

```
# Install a package from a CRAN repository.

install.packages(pkgs="dplyr")
```

Once a package is installed we can access the commands provided by that package by prefixing the command name with the package name as in `ggplot2::qplot()`. This is to say that `qplot()` is provided by the ggplot2 package.

Another example of a useful command that we will find ourselves using often is `glimpse()` from dplyr. This command can be accessed in the R console as `dplyr::glimpse()` once the dplyr package has been installed. This particular command accepts an *argument* `x=` which names the dataset we wish to glimpse. In the following R example we `dplyr::glimpse()` the weatherAUS dataset from the rattle package.

```
# Review the dataset.

dplyr::glimpse(x=rattle::weatherAUS)

## Observations: 138,307
## Variables: 24
## $ Date          <date> 2008-12-01, 2008-12-02, 2008-12-03,...
## $ Location      <fctr> Albury, Albury, Albury, Albury, Alb...
## $ MinTemp       <dbl> 13.4, 7.4, 12.9, 9.2, 17.5, 14.6, 14...
## $ MaxTemp       <dbl> 22.9, 25.1, 25.7, 28.0, 32.3, 29.7, ...
....
```

As a convention used in this book the output from running R commands is prefixed with "## ". The "#" introduces a comment in an R script file and tells R to ignore everything that follows on that line. We use the "## " convention throughout the book to clearly identify output produced by R. When we run these commands ourselves in R this prefix is not displayed.

Long lines of output are also truncated for our presentation here. The ... at the end of the lines and the .... at the end of the output indicate that the output has been truncated for the sake of keeping our printed output to an informative minimum.

We can ***attach*** a package to our *current* R session from our local *library* for added convenience. This will make the command available during this specific R session without the requirement to specify the package name each time we use the command. Attaching a package tells the R software to look within that package (and then to look within any other attached packages) when it needs to find out what the command should do (the ***definition*** of the command). Thus, we can write:

```
# Review the dataset.

glimpse(x=weatherAUS)
```

We can see which packages are currently attached using `base::search()`. The order in which they are listed here corresponds to the order in which R searches for the definition of a command.

```
base::search()

## [1] ".GlobalEnv"         "package:stats"
## [3] "package:graphics"   "package:grDevices"
## [5] "package:utils"      "package:datasets"
## [7] "package:methods"    "Autoloads"
## [9] "package:base"
```

Notice that a collection of *package*s is installed by default. We can also see a couple of other special objects called (`.GlobalEnv` and `Autoloads`).

A package is attached using the `base::library()` command which takes an argument to identify the `package=` we wish to *attach*.

```
# Load packages from the local library into the R session.

library(package=dplyr)
library(package=rattle)
```

Running these two commands will affect the search path by placing these packages early within the path.

```
base::search()

##  [1] ".GlobalEnv"        "package:rattle"
##  [3] "package:dplyr"     "package:stats"
##  [5] "package:graphics"  "package:grDevices"
##  [7] "package:utils"     "package:datasets"
##  [9] "package:methods"   "Autoloads"
## [11] "package:base"
```

By attaching the dplyr package we can drop the package name prefix for any commands from the package. Similarly by attaching rattle we can drop the package name prefix from the name of the dataset. Our previous `dplyr::glimpse()` command can be simplified to that which we saw above.

We can actually simplify this a little more. Often for a command we don't have to explicitly name all of the arguments. In the following example we drop the `package=` and the `x=` arguments from the respective commands since the commands themselves know what to expect implicitly.

```
# Load packages from the local library into the R session.

library(dplyr)
library(rattle)

# Review the dataset.

glimpse(weatherAUS)

## Observations: 138,307
## Variables: 24
## $ Date          <date> 2008-12-01, 2008-12-02, 2008-12-03,...
## $ Location      <fctr> Albury, Albury, Albury, Albury, Alb...
## $ MinTemp       <dbl> 13.4, 7.4, 12.9, 9.2, 17.5, 14.6, 14...
## $ MaxTemp       <dbl> 22.9, 25.1, 25.7, 28.0, 32.3, 29.7, ...
....
```

A number of packages are automatically attached when R starts. The first `base::search()` command above returned a vector of packages and since we had yet to attach any further packages those listed are the ones automatically attached. One of those is the base (R Core Team, 2017) package which provides the `base::library()` command.

In summary, when we interact with R we can usually drop the package prefix for those commands that can be found in one (or more) of the attached packages. Throughout the text in this book we will retain the package prefix to clarify where each command comes from. However, within the code we will tend to drop the package name prefix.

A prefix can still be useful in larger programs to ensure we are using the correct command and to communicate to the human reader where the command comes from. Some packages do implement commands with the same names as commands defined differently in other packages. The prefix notation is then essential in specifying which command we are referring to.

As noted above the following chapters will begin with a list of *packages* to *attach* from the *library* into the R session. Below is an example of attaching five common packages for our work. Attaching the listed packages will allow the examples presented within the chapter to be replicated. In the code below take note of the use of the hash (#) to introduce a comment which is ignored by R—R will not attempt to understand the comments as commands. Comments are there to assist the human reader in understanding our programs, which is a very important aspect to writing programs.

The packages that we attach are dplyr (Wickham *et al.*, 2017a), ggplot2 (Wickham and Chang, 2016), magrittr (Bache and Wickham, 2014), rattle (Williams, 2017) and readr (Wickham *et al.*, 2017b).

```
# Load packages required for this script.

library(dplyr)     # Data wrangling and glimpse().
library(ggplot2)   # Visualise data.
library(magrittr)  # Pipes %>%, %<>%, %T>%, %$%.
library(rattle)    # The weatherAUS dataset and normVarNames().
library(readr)     # Efficient reading of CSV data.
```

In starting up an R session (for example, by starting up RStudio) we can enter the above `library()` commands into an R script file created using the New R Script File menu option in RStudio and then ask RStudio to Run the commands. RStudio will send each command to the R Console which sends the command on to the R interpreter. It is the R interpreter that then runs the commands. If R responds that the package is not available, then the package will need to be installed, which we can do from RStudio's Tools menu or by directly using utils::install.packages() as we saw above. This requires an Internet conenction.

2.3 Functions, Commands and Operators

So far we have introduced the concept of R *commands* that we use to instruct the R interpreter to perform particular actions. In fact such commands are formally referred to as *functions* and we generally use this term in the same context as that of functions in mathematics. A function simply takes a set of inputs and produces an output. As we work through each chapter many new R *functions* will be introduced. We will generally identify each as a *function*, a *command* or an *operator*.

All R *functions* take arguments (a set of inputs) and return values (an output). When we run a function (rather than a *command*) we are interested in the value that it returns. Below is a simple example where we base::sum() two numbers.[*]

```
# Add two numbers.

sum(1, 2)

## [1] 3
```

As previously we precede the output from the function by the double hash (##). We will however not see the double hash in the R Console when we run the functions ourselves. Instead, the output

[*]It is a good idea to replicate all of the examples presented here in R. In RStudio simply open an R script file to edit and type the text sum(1, 2) using the editor. Then instruct RStudio to Run the command in the R Console.

we see will begin with the [1] which indicates that the returned value is a *vector* which starts counting from index 1. A ***vector*** is a collection of items which we can access through a sequential index—in this case the vector has only one item.

We can store the resulting value (the output) from running the function (the value being a vector of length 1 containing just the item 3) into a *variable.* A ***variable*** is a name that we can use to refer to a specific location in the computer's memory where we store data while our programs are running. To store data in the computer's memory so that we can later refer to that data by the specific variable name we use the ***assignment*** *operator* base::<-.

```
# Add two numbers and assign the result into a variable.

v <- sum(1, 2)
```

We can now access the value stored in the variable (or actually stored in the computer's memory that the variable name refers to) simply by requesting through the R Console that R "run" the command v. We do so below with a line containing just v. In fact, the R interpreter runs the base::print() function to display the contents of the computer's memory that v refers to—this is a convenience that the R interpreter provides for us.

```
# Print the value stored in a variable.

v

## [1] 3

print(v)

## [1] 3
```

By indexing the variable with [1] we can ask for the first item referred to by v. Noting that the variable v is a vector with only a single item when we try to index a second item we get an NA (meaning not available or a missing value).

```
# Access a particular value from a vector of data.

v[1]

## [1] 3

v[2]

## [1] NA
```

For ***commands*** (rather than *functions*) we are generally more interested in the actions or side-effects that are performed by the command rather than the value returned by the command. For example, the `base::library()` command will attach a package from the library (the action) and consequently modifies the search path that R uses to find commands (the side-effect).

```
# Load package from the local library into the R session.

library(rattle)
```

A *command* is also a function and does in fact return a value. In the above example we do not see any value printed. Some functions in R are implemented to return values invisibly. This is the case for `base::library()`. We can ask R to print the value when the returned result is invisible using `base::print()`.

```
# Demonstrate that library() returns a value invisibly.

l <- library(rattle)
print(l)

##  [1] "readr"      "extrafont" "Hmisc"      "Formula"
##  [5] "survival"   "lattice"   "magrittr"   "xtable"
##  [9] "rattle"     "ggplot2"   "tidyr"      "stringr"
## [13] "dplyr"      "knitr"     "stats"      "graphics"
## [17] "grDevices"  "utils"     "datasets"   "base"
```

We see that the value returned by `base::library()` is a vector of character strings. Each character string names a package that has been attached during this session of R. Notice that the vector begins with item [1] and the item counts are shown at the beginning of each line. We saved the resulting vector into a variable and can now index the variable to identify packages at specific locations within the vector.

```
# Load a package and save the value returned.

l <- library(rattle)

# Review one of the returned values.

l[7]

## [1] "magrittr"
```

The third type of function is the *operator*. We use ***operator*s** in the flow of an expression. Technically they are called *infix* operators.

In the following example we use the base::+ infix operator.

```
# Add two numbers using an infix operator.

1 + 2

## [1] 3
```

Internally R converts the operator into a functional prefix form to be run in the same way as any other function. We can see this effectively in the following code. We quote the function name (base::+) to avoid it being interpreted as the infix operator.

```
# Add two numbers using the equivalent functional form.

`+`(1, 2)

## [1] 3
```

The key thing to remember is that all commands and operators are functions and we instruct R to perform tasks by running functions.

2.4 Pipes

We now understand that everything in R is basically a function (or data). That is, they are actions that take a set of inputs and return an output—they are the verbs of our language for constructing sentences. As we proceed we will learn many new functions useful for the data scientist. However, as data scientists we will construct longer sentences that string verbs together to deliver the full power of functions. We combine dedicated and well designed and implemented functions to achieve something more powerful than any single function might be able to achieve on its own.

In this section we introduce the concept of *pipes* as a powerful operator for combining functions to build sentences that can deliver results from complex processing with relative simplicity.

The ***pipe*** will be familiar to those who have used the Unix and Linux operating systems. The idea is to pass the output of one function on to another function as that function's input through a sequence of steps. Each function does one task and aims to do that task very well, very accurately, and very simply from a user's point of view. We can then string together many such specialist functions to deliver very complex and quite sophisticated data transformations in an easily accessible manner. Pipes are available in R through the magrittr.

We will illustrate the concept of pipes by again making use of the rattle::weatherAUS dataset. We can review the basic contents of the rattle::weatherAUS dataset by printing it.

```
# List all columns for all observations of the dataset.

glimpse(weatherAUS)

## Observations: 138,307
## Variables: 24
## $ Date          <date> 2008-12-01, 2008-12-02, 2008-12-03,...
## $ Location      <fctr> Albury, Albury, Albury, Albury, Alb...
## $ MinTemp       <dbl> 13.4, 7.4, 12.9, 9.2, 17.5, 14.6, 14...
## $ MaxTemp       <dbl> 22.9, 25.1, 25.7, 28.0, 32.3, 29.7, ...
....
```

We might be interested in just a few variables from the dataset. For that we will dplyr::select() the variables by *piping* the dataset into the dplyr::select() function.

```
# Select columns from the dataset.

weatherAUS %>%
  select(MinTemp, MaxTemp, Rainfall, Sunshine) %>%
  glimpse()

## Observations: 138,307
## Variables: 4
## $ MinTemp  <dbl> 13.4, 7.4, 12.9, 9.2, 17.5, 14.6, 14.3, 7...
## $ MaxTemp  <dbl> 22.9, 25.1, 25.7, 28.0, 32.3, 29.7, 25.0,...
## $ Rainfall <dbl> 0.6, 0.0, 0.0, 0.0, 1.0, 0.2, 0.0, 0.0, 0...
## $ Sunshine <dbl> NA, NA, NA, NA, NA, NA, NA, NA, NA, NA, N...
....
```

Notice how **weatherAUS** by itself will list the whole of the dataset. Piping the whole dataset to dplyr::**select()** using the *pipe* operator magrittr::%>% tells R to send the rattle::**weatherAUS** dataset on the left to the dplyr::**select()** function on the right. We provided dplyr::**select()** with an argument listing the columns we wish to select. The end result returned as the output of the pipeline is a subset of the original dataset containing just the named columns.

We can easily produce a base::**summary()** of these numeric variables. We *pipe* the dataset produced by dplyr::**select()** into base::**summary()** as below.

```
# Select colums from the dataset and summarise the result.

weatherAUS %>%
  select(MinTemp, MaxTemp, Rainfall, Sunshine) %>%
  summary()

##     MinTemp          MaxTemp          Rainfall      
##  Min.   :-8.50   Min.   :-4.80   Min.   :  0.000  
##  1st Qu.: 7.60   1st Qu.:17.90   1st Qu.:  0.000  
##  Median :12.00   Median :22.60   Median :  0.000  
##  Mean   :12.16   Mean   :23.19   Mean   :  2.356  
##  3rd Qu.:16.80   3rd Qu.:28.20   3rd Qu.:  0.800  
```

```
## Max.   :33.90   Max.   :48.10   Max.    :371.000
....
```

Now suppose we would like to summarise only those observations where there is more than a little rain on the day of the observation. To do so we will `dplyr::filter()` the observations.

```
# Select specific columns and observations from the dataset.

weatherAUS %>%
  select(MinTemp, MaxTemp, Rainfall, Sunshine) %>%
  filter(Rainfall >= 1)

##      MinTemp MaxTemp Rainfall Sunshine
## 1       17.5    32.3      1.0       NA
## 2       13.1    30.1      1.4       NA
## 3       15.9    21.7      2.2       NA
## 4       15.9    18.6     15.6       NA
## 5       12.6    21.0      3.6       NA
....
```

We can see that this sequence of functions operating on the original `rattle::weatherAUS` dataset returns a subset of that dataset where all observations have some rain.

We saw earlier in this chapter the assignment operator `base::<-` which is used to save a result into the computer's memory and give it a name that we can refer to later. We can use this operator to save the result of the sequence of operations from the pipe into the computer's memory.

```
# Select columns/observations and save the result.

rainy_days <-
  weatherAUS %>%
  select(MinTemp, MaxTemp, Rainfall, Sunshine) %>%
  filter(Rainfall >= 1)
```

An alternative that makes logical sense within the pipe paradigm is to use the forward assignment operator `base::->` provided by R to save the resulting data into a variable but including it within the logical flow through which we think about the operations being performed.

```
# Demonstrate use of the forward assignment operator.

weatherAUS %>%
  select(MinTemp, MaxTemp, Rainfall, Sunshine) %>%
  filter(Rainfall >= 1) ->
rainy_days
```

Traditionally in R we have avoided the use of the forward assignment operator but here it makes sense. Notice in the above code that we un-indent the variable name to highlight the important side effect of the series of commands—the assignment to `rainy_days`. Logically it is clear in this sequence of commands that we begin with a dataset, operate on it through a variety of functions, and save the result into a variable.

Continuing with our pipeline example, we might want a `base::summary()` of the resulting dataset.

```
# Summarise subset of variables for observations with rainfall.

weatherAUS %>%
  select(MinTemp, MaxTemp, Rainfall, Sunshine) %>%
  filter(Rainfall >= 1) %>%
  summary()

##      MinTemp          MaxTemp          Rainfall
##  Min.   :-8.5   Min.   :-4.80   Min.   :  1.000
##  1st Qu.: 8.4   1st Qu.:15.50   1st Qu.:  2.200
##  Median :12.2   Median :19.20   Median :  4.800
##  Mean   :12.7   Mean   :20.12   Mean   :  9.739
##  3rd Qu.:17.1   3rd Qu.:24.30   3rd Qu.: 11.200
##  Max.   :28.3   Max.   :46.30   Max.   :371.000
....
```

It could be useful to contrast this with a `base::summary()` of those observations where there was virtually no rain.

```
# Summarise observations with little or no rainfall.

weatherAUS %>%
  select(MinTemp, MaxTemp, Rainfall, Sunshine) %>%
  filter(Rainfall < 1) %>%
  summary()
```

```
##      MinTemp          MaxTemp          Rainfall
##  Min.   :-8.20   Min.   :-2.10   Min.   :0.00000
##  1st Qu.: 7.30   1st Qu.:18.90   1st Qu.:0.00000
##  Median :11.90   Median :23.70   Median :0.00000
##  Mean   :11.99   Mean   :24.16   Mean   :0.06044
##  3rd Qu.:16.70   3rd Qu.:29.20   3rd Qu.:0.00000
##  Max.   :33.90   Max.   :48.10   Max.   :0.90000
....
```

Any number of functions can be included in a ***pipeline*** to achieve the results we desire. In the following chapters we will see some examples which string together 10 or more functions. Each step along the way is generally simple to understand in and of itself. The power is in what we can achieve by stringing together many simple steps to produce something quite complex.

For the technically minded we note that what is actually happening here is that a new syntax is introduced in order to increase the ease with which we humans can read the code. This is an important goal as we need to always keep in mind that we write our code for others (and ourselves later on) to read.

The above example of a pipeline is actually translated by R into the functional construct we write below. For many of us it will take quite a bit of effort to parse this traditional functional form into something we could understand. The pipeline alternative provides a clearer narrative.

```
# Functional form equivalent to the pipeline above.

summary(filter(select(weatherAUS,
                      MinTemp, MaxTemp, Rainfall, Sunshine),
               Rainfall < 1))
```


```
##      MinTemp          MaxTemp          Rainfall
##  Min.   :-8.20   Min.   :-2.10   Min.   :0.00000
##  1st Qu.: 7.30   1st Qu.:18.90   1st Qu.:0.00000
##  Median :11.90   Median :23.70   Median :0.00000
##  Mean   :11.99   Mean   :24.16   Mean   :0.06044
##  3rd Qu.:16.70   3rd Qu.:29.20   3rd Qu.:0.00000
##  Max.   :33.90   Max.   :48.10   Max.   :0.90000
....
```

Anything that improves the readability of our code is useful. Indeed we allow the computer to do the hard work of transforming a simpler sentence into this much more complex looking sentence.

There are several variations of the pipe operator available. A particularly handy operator is the assignment pipe magrittr::%<>%. This operator should be the left-most pipe of any sequence of pipes. In addition to piping the dataset on the left into the function on the right the result coming out of the right-hand pipeline is piped back to the original variable. Thus, we overwrite the original contents in memory with the results from the pipeline.

A simple example is to replace a dataset with the same data after removing some observations (rows) and variables (columns). In the example below we dplyr::**filter()** and dplyr::**select()** the dataset to reduce it to just those observations and variables of interest. The result is piped backwards to the original dataset and thus overwrites the original data (which may or may not be a good thing). We do this on a temporary copy of the dataset and use the base::**dim()** function to report on the dimensions (rows and columns) of the resulting datasets.

```
# Copy the dataset into the variable ds.

ds <- weatherAUS

# Report on the dimensions of the dataset.

dim(ds)

## [1] 138307     24

# Demonstrate an assignment pipeline.

ds %<>%
  filter(Rainfall==0) %>%
  select(MinTemp, MaxTemp, Sunshine)

# Confirm that the dataset has changed.

dim(ds)

## [1] 86311     3
```

Once again this is so-called *syntactic sugar*. The functions are effectively translated by the computer into the following code.

```
# Functional form equivalent to the pipeline above.

ds <- select(filter(weatherAUS, Rainfall==0),
             MinTemp, MaxTemp, Sunshine)
```

Another useful operation is the tee-pipe magrittr::%T>% which causes the function that follows to be run as a side-pipe whilst piping the same data into that function and also into the next function. The output from the function immediately following the tee-pipe operator is ignored. A common use case is to base::print() the result of some data processing steps whilst continuing on to assign the dataset itself to a variable. We will often see the following example.

```
# Demonstrate usage of a tee-pipe.

weatherAUS %>%
  filter(Rainfall==0) %T>%
  {head(.) %>% print()} ->
no_rain

##          Date Location MinTemp MaxTemp Rainfall Evaporation
## 1 2008-12-02   Albury     7.4    25.1        0          NA
## 2 2008-12-03   Albury    12.9    25.7        0          NA
## 3 2008-12-04   Albury     9.2    28.0        0          NA
## 4 2008-12-07   Albury    14.3    25.0        0          NA
## 5 2008-12-08   Albury     7.7    26.7        0          NA
....
```

In this simple example we could have called utils::head() with the variable no_rain after the assignment but for more complex pipes there is some elegance in including intermediate output as part of the single pipeline.

```
# Alternative to using a tee-pipe for a simple pipeline.

weatherAUS %>%
  filter(Rainfall==0) ->
no_rain
```

```
head(no_rain)

##         Date Location MinTemp MaxTemp Rainfall Evaporation
## 1 2008-12-02   Albury     7.4    25.1        0          NA
## 2 2008-12-03   Albury    12.9    25.7        0          NA
## 3 2008-12-04   Albury     9.2    28.0        0          NA
## 4 2008-12-07   Albury    14.3    25.0        0          NA
## 5 2008-12-08   Albury     7.7    26.7        0          NA
....
```

We can also include within the same pipeline a number of operations that describe the data as it is passing through.

```
# Multiple tee-pipes in a single pipeline.

weatherAUS %>%
  filter(Rainfall==0) %T>%
  {dim(.) %>% comcat()} %T>%
  {head(.) %>% print()} ->
no_rain

## 86,311 24
##         Date Location MinTemp MaxTemp Rainfall Evaporation
## 1 2008-12-02   Albury     7.4    25.1        0          NA
## 2 2008-12-03   Albury    12.9    25.7        0          NA
## 3 2008-12-04   Albury     9.2    28.0        0          NA
## 4 2008-12-07   Albury    14.3    25.0        0          NA
....
```

Here we dplyr::**filter()** the dataset to retain only those observations for which the recorded rainfall is zero. Using a tee-pipe this subset is piped into two branches. The first is bracketed to allow a sequence of functions to be composed through another pipeline. The first in the sequence of just two functions obtains the dimensions of the subset calling upon base::**dim()** to do so. The parameter to base::**dim()** is a period (.) which is used to indicate the piped dataset within the bracketed sequence. The result of base::**dim()** is passed on to rattle::**comcat()** which simply formats the incoming data to be more easily readable.

The second branch is also introduced by a tee-pipe so that effectively we have a three-way flow of our data subset. The second sequence takes the utils::**head()** of the subset and prints that. The same subset is then piped through the forward assignment on to the variable **no_rain** so as to store the subset in memory.

We conclude our introduction to pipes with a more complex example of a pipeline that processes a dataset and feeds the output into a plot command which itself is built by adding layers to the plot.

```
# Identify cities of interest.

cities <- c("Canberra", "Darwin", "Melbourne", "Sydney")

# Build the required dataset and plot it.

weatherAUS %>%
  filter(Location %in% cities) %>%
  filter(Temp3pm %>% is.na() %>% not()) %>%
  ggplot(aes(x=Temp3pm, colour=Location, fill=Location)) +
  geom_density(alpha=0.55) +
  labs(title="Distribution of the Daily Temperature at 3pm",
       subtitle="Selective Locations",
       caption="Source: Australian Bureau of Meteorology",
       x="Temperature Recorded at 3pm",
       y="Density")
```

In the example once again we use the rattle::**weatherAUS** dataset and dplyr::**filter()** it to include those observations from just four cities in Australia. We also dplyr::**filter()** out those observations that have missing values for the variable **Temp3pm**. To do this we use an embedded pipeline. The embedded pipeline pipes the **Temp3pm** data through the base::**is.na()** function which tests if the value is missing. These results are then piped to magrittr::**not()** which inverts the true/false values so that we include those that are not missing.

The plot is generated using ggplot2::**ggplot()** into which we pipe the processed dataset. We add a geometric layer using ggplot2::**geom_density()** which constructs a density plot with transparency specified through the alpha= argument. We also add a title and label the axes using ggplot2::**labs()**. The code produces the plot shown in Figure 2.4.

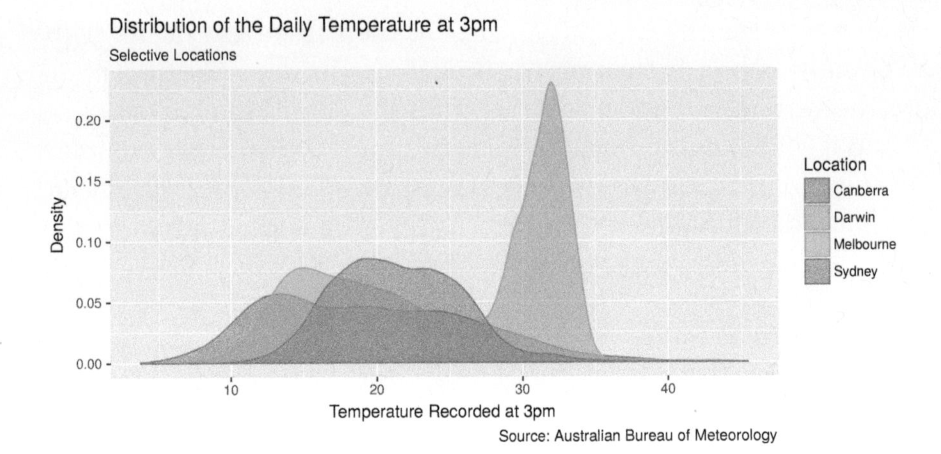

Figure 2.4: *A comparison of the distribution of the daily temperature at 3pm across 4 Australian cities.*

With a few simple steps (and recognising we have introduced quite a few new functions without explanation) we have produced a relatively sophisticated plot.

As a data scientist we would now observe and tell a story from this plot. Our narrative will begin with the observation that Darwin has quite a different and warmer pattern of temperatures at 3pm as compared to Canberra, Melbourne and Sydney. Of course, the geography of Australia would inform the logic of this observation.

2.5 Getting Help

A key skill of any programmer, including those programming over data, is the ability to identify how to access the full power of our tools. The breadth and depth of the capabilities of R means that there is much to learn around both the basics of R programming and the multitude of packages that support the data scientist.

R provides extensive in-built documentation with many introductory manuals and resources available through the RStudio `Help` tab. These are a good adjunct to our very brief introduction here to R. Further we can use RStudio's search facility for documentation on any action and we will find manual pages that provide an understanding of the purpose, arguments, and return value of *functions*, *commands*, and *operators*. We can also ask for help using the utils::? operator as in:

```
# Ask for documentation on using the library command.

? library
```

The utils package which provides this operator is another package that is attached by default to R. Thus, we can drop its prefix as we did here when we run the command in the R Console.

RStudio provides a search box which can be used to find specific topics within the vast collection of R documentation. Now would be a good time to check the documentation for the base::library() command.

2.6 Exercises

Exercise 2.1 Exploring RStudio

Install R and RStudio onto your own computer. Start RStudio and work through the Introduction to R manual, repeating the sample R codes for yourself within the R Console of RStudio, being sure to save all of the code to file. Document and layout the code according to the Style guide in Chapter 11. Optional: complete the exercise using either R markdown or KnitR.

Exercise 2.2 Documentation and Cleansing

Attach the rattle package in RStudio. Obtain a list of the functions available from the rattle package. Review the documentation for rattle::normVarNames(). Using one of the open source datasets identified in the exercises in Chapter 1, load the dataset into R and apply rattle::normVarNames() to the base::names() of the variables within the dataset. Save the code to an R script file. Optional: complete the exercise using either R markdown or KnitR.

Exercise 2.3 Interacting with RStudio

1. Create a new R script file to explore a number of visualisations of various distributions of pairs of variables from the `rattle::weatherAUS` dataset. Include at least a plot of the distribution of `WindSpeed9am` against `Humidity9am`.

2. Filter the dataset by different locations and review and compare the plots.

3. Write a short report of the `rattle::weatherAUS` dataset based on the visualisations you produce. The report should also describe the dataset, including a summary of the number of variables and observations. Describe each variable and summarise its distribution. Include the more interesting or relevant plots in your report.

Exercise 2.4 Telling a narrative.

Load the `rattle::weatherAUS` dataset into R. Explore the rainfall for a variety of locations, generating different plots similar to the two examples provided in this chapter. Can you determine any relationship between the values of the variable `RainTomorrow` and the other variables. Write a short report of your discoveries including the R code you develop and sample plots that support your narrative.

3

Data Wrangling

When we begin with a new data science project we will begin with understanding the business problem to be tackled. This includes ensuring all participants in the project understand the goals, the success criteria, and how the results will be deployed into production in the business. We then liaise with the business data technicians to identify the available data. This is followed by a data phase where we work with the business data technicians to access and ingest the data into R. We are then in a position to move on with our journey to the discovery of new insights driven by the data. By *living and breathing* the data in the context of the business problem we gain our bearings and feed our intuitions as we journey.

In this chapter we present and then capture a common series of steps that we follow as the data phase of data science. As we progress through the chapter we will build a ***template***[*] designed to be reused for journeys through other datasets. As we foreshadowed in Chapter 2 rather than delving into the intricacies of the R language we immerse ourselves into using R to achieve our outcomes, learning more about R as we proceed.

As we will for each chapter we begin with loading the packages required for this chapter. Packages used in this chapter include FSelector (Romanski and Kotthoff, 2016), dplyr (Wickham *et al.*, 2017a), ggplot2 (Wickham and Chang, 2016), lubridate (Grolemund *et al.*, 2016), randomForest (Breiman *et al.*, 2015), rattle (Williams, 2017), readr (Wickham *et al.*, 2017b), scales (Wickham, 2016), stringi (Gagolewski *et al.*, 2017), stringr (Wick-

[*]The template will consist of programming code that can be reused with little or no modification on a new dataset. The intention is that to get started with a new dataset only a few lines at the top of the template need to be modified.

ham, 2017a), tibble (Müller and Wickham, 2017), tidyr (Wickham, 2017b) and magrittr (Bache and Wickham, 2014).

```
# Load required packages from local library into the R session.

library(FSelector)     # Feature selection: information.gain().
library(dplyr)         # Data wrangling, glimpse() and tbl_df().
library(ggplot2)       # Visualise data.
library(lubridate)     # Dates and time.
library(randomForest) # Impute missing values with na.roughfix().
library(rattle)        # weatherAUS data and normVarNames().
library(readr)         # Efficient reading of CSV data.
library(scales)        # Format comma().
library(stringi)       # String concat operator %s+%.
library(stringr)       # String operations.
library(tibble)        # Convert row names into a column.
library(tidyr)         # Prepare a tidy dataset, gather().
library(magrittr)      # Pipes %>%, %T>% and equals(), extract().
```

3.1 Data Ingestion

To begin the data phase of a project we typically need to ingest data into R. For our purposes we will ingest the data from the simplest of sources—a text-based CSV (comma separate value) file. Practically any source format is supported by R through the many packages available. We can find, for example, support for ingesting data directly from Excel spreadsheets or from database servers.

The **weatherAUS** dataset from rattle will serve to illustrate the preparation of data for modelling. Both a CSV file and an R dataset are provided by the package and the dataset is also available directly from the Internet on the rattle web site. We will work with the CSV file as the typical pathway for loading data into R.

Identifying the Data Source

We first identify and record the location of the CSV file to analyse. R is capable of loading data directly from the Internet and so we will illustrate how to load the CSV file from the rattle web site itself. The location of the file (the so-called URL or universal resource locator) will be saved as a string of characters in a variable called dspath—the path to the dataset. This is achieved through the following assignment which we type into our R script file within RStudio. The command is then executed in RStudio by clicking the Run button whilst the cursor is located on the line of code within the script file.

```
# Idenitfy the source location of the dataset.

dspath <- "http://rattle.togaware.com/weatherAUS.csv"
```

The assignment operator <- will store the value on the right-hand side (the string of characters enclosed within quotation marks) into the computer's memory and we can later refer to it as the R variable dspath—that is, we can retrieve the string simply by reference to the variable dspath.

Reading the Data

Having identified the source of the dataset we can read the dataset into the memory of the computer using readr::read_csv(). This function returns a ***data frame*** (though it is actually an enhanced data frame known as a ***table data frame***) which is the basic data structure used to store a dataset within R. The data is stored as a table consisting of rows (***observations***) and columns (***variables***). We store the dataset (as a data frame) in the computer's memory and reference it by the R variable weatherAUS. It will then be ready to process using R.

```
# Ingest the dataset.

weatherAUS <- read_csv(file=dspath)

## Parsed with column specification:
```

```
## cols(
##   .default = col_character(),
##   Date = col_date(format = ""),
##   MinTemp = col_double(),
##   MaxTemp = col_double(),
##   Rainfall = col_double(),
##   WindGustSpeed = col_integer(),
##   WindSpeed9am = col_integer(),
##   WindSpeed3pm = col_integer(),
##   Humidity9am = col_integer(),
##   Humidity3pm = col_integer(),
##   Pressure9am = col_double(),
##   Pressure3pm = col_double(),
##   Cloud9am = col_integer(),
##   Cloud3pm = col_integer(),
##   Temp9am = col_double(),
##   Temp3pm = col_double(),
##   RISK_MM = col_double()
## )
## See spec(...) for full column specifications.
```

Template Variables

To support our goal of creating a reusable template we create a reference to the original dataset using a template (or generic) variable. The new variable will be called `ds` (short for dataset).

```
# Take a copy of the dataset into a generic variable.

ds <- weatherAUS
```

Both `ds` and `weatherAUS` will initially reference the same dataset within the computer's memory. As we modify `ds` those modifications will only affect the data referenced by `ds` and not the data referenced by `weatherAUS`. Effectively an extra copy of the dataset in the computer's memory will start to grow as we change the data from its original form.* From here on we no longer refer

*R avoids making copies of datasets unnecessarily and so a simple assignment does not create a new copy. As modifications are made to one or the other copy of a dataset then extra memory will be used to store the columns that differ between the datasets.

to the dataset as **weatherAUS** but as **ds**. This allows the following steps to be generic—turning the R code into a *template* requiring only minor modification when used with a different dataset assigned into **ds**. Often we will find that we can simply load a different dataset into memory, store it as **ds** and the following code remains essentially unchanged.

The next few steps of our template record the name of the dataset and a generic reference to the dataset.

```
# Prepare the dataset for usage with our template.

dsname <- "weatherAUS"
ds     <- get(dsname)
```

We are a little tricky here in recording the dataset name in the variable **dsname** and then using the function base::**get()** to make a copy of the original dataset reference and to link it to the generic variable **ds**. We could simply assign the data to **ds** directly as we saw above. Either way the generic variable **ds** refers to the same dataset. The use of base::**get()** allows us to be generic within the template.

The use of generic variables within a template for the tasks we perform on each new dataset will have obvious advantages but we need to be careful. A disadvantage is that we may be working with several datasets and accidentally overwrite previously processed datasets referenced using the same generic variable (**ds**). The processing of the dataset might take some time and so accidentally losing it is not an attractive proposition. Be careful to manage the naming of datasets appropriately to avoid any loss of processed data.

The Shape of the Data

Once the dataset is loaded we seek a basic understanding of the data—its shape. We are interested in the size of the dataset in terms of the number of observations (rows) and variables (columns). We can simply type the variable name that stores the dataset and will be presented with a summary of the actual contents of the dataset.

```
# Print the dataset.

ds

## # A tibble: 138,307 x 24
##          Date Location MinTemp MaxTemp Rainfall Evaporation
##        <date>    <chr>   <dbl>   <dbl>    <dbl>       <chr>
##  1 2008-12-01   Albury    13.4    22.9      0.6        <NA>
##  2 2008-12-02   Albury     7.4    25.1      0.0        <NA>
##  3 2008-12-03   Albury    12.9    25.7      0.0        <NA>
##  4 2008-12-04   Albury     9.2    28.0      0.0        <NA>
##  5 2008-12-05   Albury    17.5    32.3      1.0        <NA>
##  6 2008-12-06   Albury    14.6    29.7      0.2        <NA>
##  7 2008-12-07   Albury    14.3    25.0      0.0        <NA>
##  8 2008-12-08   Albury     7.7    26.7      0.0        <NA>
##  9 2008-12-09   Albury     9.7    31.9      0.0        <NA>
## 10 2008-12-10   Albury    13.1    30.1      1.4        <NA>
## # ... with 138,297 more rows, and 18 more variables:
## #   Sunshine <chr>, WindGustDir <chr>, WindGustSpeed <int>,
## #   WindDir9am <chr>, WindDir3pm <chr>, WindSpeed9am <int>,
## #   WindSpeed3pm <int>, Humidity9am <int>, Humidity3pm <int>,
## #   Pressure9am <dbl>, Pressure3pm <dbl>, Cloud9am <int>,
## #   Cloud3pm <int>, Temp9am <dbl>, Temp3pm <dbl>,
## #   RainToday <chr>, RISK_MM <dbl>, RainTomorrow <chr>
```

The function base::**dim()** will provide dimension information (observations and variables) as will base::**nrow()** and base::**ncol()**. We use rattle::**comcat()** to format numbers with commas.

```
# Basic size information.

dim(ds)  %>% comcat()

## 138,307 24

nrow(ds) %>% comcat()

## 138,307

ncol(ds) %>% comcat()

## 24
```

A useful alternative for our initial insight into the dataset is to use `tibble::glimpse()`.

```
# A quick view of the contents of the dataset.

glimpse(ds)

## Observations: 138,307
## Variables: 24
## $ Date          <date> 2008-12-01, 2008-12-02, 2008-12-03,...
## $ Location      <chr> "Albury", "Albury", "Albury", "Albur...
## $ MinTemp       <dbl> 13.4, 7.4, 12.9, 9.2, 17.5, 14.6, 14...
## $ MaxTemp       <dbl> 22.9, 25.1, 25.7, 28.0, 32.3, 29.7, ...
## $ Rainfall      <dbl> 0.6, 0.0, 0.0, 0.0, 1.0, 0.2, 0.0, 0...
## $ Evaporation   <chr> NA, NA, NA, NA, NA, NA, NA, NA, NA, ...
## $ Sunshine      <chr> NA, NA, NA, NA, NA, NA, NA, NA, NA, ...
## $ WindGustDir   <chr> "W", "WNW", "WSW", "NE", "W", "WNW",...
## $ WindGustSpeed <int> 44, 44, 46, 24, 41, 56, 50, 35, 80, ...
## $ WindDir9am    <chr> "W", "NNW", "W", "SE", "ENE", "W", "...
## $ WindDir3pm    <chr> "WNW", "WSW", "WSW", "E", "NW", "W",...
## $ WindSpeed9am  <int> 20, 4, 19, 11, 7, 19, 20, 6, 7, 15, ...
## $ WindSpeed3pm  <int> 24, 22, 26, 9, 20, 24, 24, 17, 28, 1...
## $ Humidity9am   <int> 71, 44, 38, 45, 82, 55, 49, 48, 42, ...
## $ Humidity3pm   <int> 22, 25, 30, 16, 33, 23, 19, 19, 9, 2...
## $ Pressure9am   <dbl> 1007.7, 1010.6, 1007.6, 1017.6, 1010...
## $ Pressure3pm   <dbl> 1007.1, 1007.8, 1008.7, 1012.8, 1006...
## $ Cloud9am      <int> 8, NA, NA, NA, 7, NA, 1, NA, NA, NA,...
## $ Cloud3pm      <int> NA, NA, 2, NA, 8, NA, NA, NA, NA, NA...
## $ Temp9am       <dbl> 16.9, 17.2, 21.0, 18.1, 17.8, 20.6, ...
## $ Temp3pm       <dbl> 21.8, 24.3, 23.2, 26.5, 29.7, 28.9, ...
## $ RainToday     <chr> "No", "No", "No", "No", "No", "No", ...
## $ RISK_MM       <dbl> 0.0, 0.0, 0.0, 1.0, 0.2, 0.0, 0.0, 0...
## $ RainTomorrow  <chr> "No", "No", "No", "No", "No", "No", ...
```

Normalizing Variable Names

Next we review the variables included in the dataset. The function `base::names()` will list the names of the variables (columns).

```
# Identify the variables of the dataset.

names(ds)

##  [1] "Date"           "Location"       "MinTemp"
```

```
## [4] "MaxTemp"        "Rainfall"      "Evaporation"
## [7] "Sunshine"       "WindGustDir"   "WindGustSpeed"
## [10] "WindDir9am"    "WindDir3pm"    "WindSpeed9am"
## [13] "WindSpeed3pm"  "Humidity9am"   "Humidity3pm"
## [16] "Pressure9am"   "Pressure3pm"   "Cloud9am"
....
```

The names of the variables within the dataset as supplied to us may not be in any particular standard form and may use different conventions. For example, we might see a mix of upper and lower case letters (`WindSpeed9AM`) or variable names that are very long (`Wind_Speed_Recorded_Today_9am`) or use sequential numbers to identify each variable (`V004` or `V010_wind_speed`) or use codes (`XVn34_windSpeed`) or any number of other conventions. Often we prefer and it is convenient to simplify the variable names to ease our processing and to enforce a standard and consistent naming convention for ourselves. This will help us in interacting the data without regular reference to how the variables are named.

A useful convention is to map all variable names to lowercase. R is case sensitive so that doing this will result in different variable names as far as R is concerned. Such normalisation is useful when different upper/lower case conventions are intermixed inconsistently in names like `Incm_tax_PyBl`. Remembering how to capitalize when interactively exploring the data with thousands of such variables can be quite a cognitive load. Yet we often see such variable names arising in practise especially when we import data from databases which are often case insensitive.

We use `rattle::normVarNames()` to make a reasonable attempt of converting variables from a dataset into a preferred standard form. The actual form follows the style we introduce in Chapter 11. The following example shows the original names and how they are transformed into a normalized form. Here we make extensive use of the function `base::names()` to work with the variable names.*

*When the name of a variable within a dataset is changed a new copy of that value of that variable is created so that now `ds` and `weatherAUS` will now be referring to different datasets in memory.

```
# Review the variables before normalising their names.

names(ds)

##  [1] "Date"          "Location"      "MinTemp"
##  [4] "MaxTemp"       "Rainfall"      "Evaporation"
##  [7] "Sunshine"      "WindGustDir"   "WindGustSpeed"
## [10] "WindDir9am"    "WindDir3pm"    "WindSpeed9am"
## [13] "WindSpeed3pm"  "Humidity9am"   "Humidity3pm"
## [16] "Pressure9am"   "Pressure3pm"   "Cloud9am"
....

# Normalise the variable names.

names(ds) %<>% normVarNames()

# Confirm the results are as expected.

names(ds)

##  [1] "date"           "location"       "min_temp"
##  [4] "max_temp"       "rainfall"       "evaporation"
##  [7] "sunshine"       "wind_gust_dir"  "wind_gust_speed"
## [10] "wind_dir_9am"   "wind_dir_3pm"   "wind_speed_9am"
## [13] "wind_speed_3pm" "humidity_9am"   "humidity_3pm"
## [16] "pressure_9am"   "pressure_3pm"   "cloud_9am"
....
```

Notice the use of the assignment pipe here as introduced in Chapter 2. Recall that the `magrittr::%<>%` operator will pipe the left-hand data to the function on the right-hand side and then return the result to the left-hand side overwriting the original contents of the memory referred to on the left-hand side. In this case, the left-hand side refers to the variable names of the dataset.

3.2 Data Review

Once we have loaded the dataset and had our initial review of its size and cleaned up the variable names the next step is to understand its structure—that is, understand what the data within the dataset looks like.

Structure

A basic summary of the structure of the dataset is presented using `tibble::glimpse()` as we saw above.

```
# Review the dataset.

glimpse(ds)

## Observations: 138,307
## Variables: 24
## $ date          <date> 2008-12-01, 2008-12-02, 2008-12-0...
## $ location      <chr> "Albury", "Albury", "Albury", "Alb...
## $ min_temp      <dbl> 13.4, 7.4, 12.9, 9.2, 17.5, 14.6, ...
## $ max_temp      <dbl> 22.9, 25.1, 25.7, 28.0, 32.3, 29.7...
## $ rainfall      <dbl> 0.6, 0.0, 0.0, 0.0, 1.0, 0.2, 0.0,...
## $ evaporation   <chr> NA, NA, NA, NA, NA, NA, NA, NA, NA...
....
```

From this summary we see the variable names, their data types and the first few values of the variable. We can see a variety of data types here, ranging from **Date** (`date`), through **character** (`chr`) and **numeric** (`dbl`).

We confirm the data looks as we would expect and begin to gain some insight into the data itself. We might start asking questions such as whether the `date` values are a sequence of days as we might expect. The first few `locations` are listed as `Albury` and so we might ask what the other values are. We see minimum and maximum temperatures and we note the rainfall and evaporation. We expect each of these to be numeric though we observe that evaporation is reported as a character variable (which we will come back to later). For the sample above we only see missing values for evaporation. For the numerics we will want to understand the distributions of the values of the variables.

Contents

Generally our datasets are very large, with many observations (often in the millions) and many variables (sometimes in the thousands). We can't be expected to browse through all of the observations and variables. Instead we can review the contents of the dataset using `utils::head()` and `utils::tail()` to review the top six (by default) and the bottom six observations.

```
# Review the first few observations.

head(ds)

## # A tibble: 6 x 24
##         date location min_temp max_temp rainfall evaporation
##       <date>    <chr>    <dbl>    <dbl>    <dbl>       <chr>
## 1 2008-12-01   Albury     13.4     22.9      0.6        <NA>
## 2 2008-12-02   Albury      7.4     25.1      0.0        <NA>
## 3 2008-12-03   Albury     12.9     25.7      0.0        <NA>
## 4 2008-12-04   Albury      9.2     28.0      0.0        <NA>
....

# Review the last few observations.

tail(ds)

## # A tibble: 6 x 24
##         date location min_temp max_temp rainfall evaporation
##       <date>    <chr>    <dbl>    <dbl>    <dbl>       <chr>
## 1 2017-01-25    Uluru     21.3     34.2      0.2        <NA>
## 2 2017-01-26    Uluru     23.6     35.5      0.0        <NA>
## 3 2017-01-27    Uluru     24.6     37.9      4.2        <NA>
## 4 2017-01-28    Uluru     24.7     39.2      0.2        <NA>
....
```

We can also have a look at some random observations from the dataset to provide a little more insight. Here we use `dplyr::sample_n()` to randomly select 6 rows from the dataset.

```
set.seed(2)
```

```
# Review a random sample of observations.

sample_n(ds, size=6)

## # A tibble: 6 x 24
##         date      location min_temp max_temp rainfall
##       <date>         <chr>    <dbl>    <dbl>    <dbl>
## 1 2016-04-22       Penrith     13.8     26.8      0.0
## 2 2015-11-10 MountGambier     12.6     19.1      0.8
## 3 2015-05-15     Dartmoor      9.1     13.6      1.4
## 4 2009-09-08       Penrith      6.4     22.4      0.2
....
```

All the time we are building a picture of the data we are looking at. We note the `date` appears to be a daily sequence starting from December 2008 and ending in 2017. We also note that `evaporation` is often but not always missing.

3.3 Data Cleaning

Identifying Factors

On loading the dataset into R we can see that a few variables have been identified as having **character** (string of characters) as the values they store. Such variables are often called *categoric* variables. Within R these are usually represented as a data type called **factor** and handled specially by many of the modelling algorithms. Where the **character** data takes on a limited number of possible values we will convert the variable from **character** into **factor** (categoric) so as to take advantage of some special handling.

A **factor** is a variable that can only take on a specific number of known distinct values which we call the ***levels*** of the **factor**. For datasets that we load into R we will not always have examples of all levels of a factor. Consequently it is not always possible to automatically list all of the levels required for the definition of a factor. Thus we load these variables by default as **character** and then convert them to **factor** as required.

From our review of the data so far we start to make some observations about the character variables. The first is `location`. We note that several locations were reported in the above exploration of the dataset. We can confirm the number of locations by counting the number of `base::unique()` values the variable has in the original dataset.

```
# How many locations are represented in the dataset.

ds$location %>%
  unique() %>%
  length()

## [1] 49
```

We may not know in general what other locations we will come across in related datasets and we already have quite a collection of 49 locations. We might normally decide to retain this variable as a character data type, but for illustrative purposes we will convert it to a factor.

```
ds$location %<>% as.factor()
```

We next review a `base::table()` of the distribution of the locations.

```
table(ds$location)

##
##         Adelaide           Albany            Albury
##             3047             2894              2894
##     AliceSprings    BadgerysCreek          Ballarat
##             2894             2863              2894
##          Bendigo         Brisbane            Cairns
....
```

Two related variables that are class character that might be better represented as factors are `rain_today` and `rain_tomorrow`. We can review the distribution of their values with the following code.

```
# Review the distribution of observations across levels.

ds %>%
  select(starts_with("rain_")) %>%
  sapply(table)

##     rain_today rain_tomorrow
## No      104498        104493
## Yes      30279         30283
```

Here we dplyr::select() from the dataset those variables that start with the string rain_ and then build a base::table() over those variables in the subset of the original dataset selected. We use base::sapply() to apply base::table() to the selected columns since the function takes a single column from the dataset as its argument. This function counts the frequency of the occurence of each value of a variable within the dataset.

We confirm that No and Yes are the only values these two variables have and so it makes sense to convert them both to factors. We will keep the ordering as alphabetic and so a simple call to base::factor() will convert each variable from character to factor using base::lapply(). Note the use of base::data.frame() and dplyr::tbl_df() to ensure the data is in the correct form to overwrite the original columns in the dataset.

```
# Note the names of the rain variables.

ds %>%
  select(starts_with("rain_")) %>%
  names() %T>%
  print() ->
vnames

## [1] "rain_today"    "rain_tomorrow"

# Confirm these are currently character variables.

ds[vnames] %>% sapply(class)

##    rain_today rain_tomorrow
##   "character"   "character"
```

```
# Convert these variables from character to factor and confirm.

ds[vnames] %<>%
  lapply(factor) %>%
  data.frame() %>%
  tbl_df() %T>%
  {sapply(., class) %>% print()}

##     rain_today rain_tomorrow
##       "factor"      "factor"
```

We can again obtain a distribution of the variables to confirm that all we have changed is the data type.

```
# Verify the distribution has not changed.

ds %>%
  select(starts_with("rain_")) %>%
  sapply(table)

##     rain_today rain_tomorrow
## No      104498        104493
## Yes      30279         30283
```

The three wind direction variables identified as `wind_gust_dir`, `wind_dir_9am` and `wind_dir_3pm` are also **character** variables. We will want to review their distribution of values and can do so in a similar way. Here we `dplyr::select()` from the dataset those variables containing the string `_dir` and then build a `base::table()` over those variables in the selected subset of the original dataset. We again use `base::sapply()` with `base::table()` to count the frequency of the occurence of each level of the factors within the dataset.

```
# Review the distribution of observations across levels.

ds %>%
  select(contains("_dir")) %>%
  sapply(table)

##     wind_gust_dir wind_dir_9am wind_dir_3pm
## E            8628         8622         8024
```

```
## ENE        7669      7436      7456
## ESE        6894      7157      8000
## N          8922     11339      8520
## NE         6822      7303      7938
## NNE        6260      7763      6332
## NNW        6301      7680      7516
## NW         7791      8404      8306
## S          8670      8241      9359
## SE         8802      8751     10227
## SSE        8707      8573      8857
## SSW        8130      7199      7572
## SW         8554      7930      8878
## W          9577      8128      9733
## WNW        7987      7112      8563
## WSW        8722      6689      9142
```

From the table we notice 16 compass directions. All compass directions are represented and so we will convert these character variables into factors. Notice that the values of the variables are listed in alphabetic order in the above and a simple conversion to a factor will retain the alphabetic order. We might know however that the compass orders the directions in a well-defined manner (from `N`, `NNE`, to `NW` and `NNW`). With this knowledge we will force the levels to have the appropriate ordering and also let `base::factor()` know that the levels are ordered with `ordered=TRUE`.

```
# Levels of wind direction are ordered compass directions.

compass <- c("N", "NNE", "NE", "ENE",
             "E", "ESE", "SE", "SSE",
             "S", "SSW", "SW", "WSW",
             "W", "WNW", "NW", "NNW")

# Note the names of the wind direction variables.

ds %>%
  select(contains("_dir")) %>%
  names() %T>%
  print() ->
vnames

## [1] "wind_gust_dir" "wind_dir_9am"  "wind_dir_3pm"
```

```
# Confirm these are currently character variables.

ds[vnames] %>% sapply(class)

## wind_gust_dir  wind_dir_9am  wind_dir_3pm
##   "character"   "character"   "character"

# Convert these variables from character to factor and confirm.

ds[vnames] %<>%
  lapply(factor, levels=compass, ordered=TRUE) %>%
  data.frame() %>%
  tbl_df() %T>%
  {sapply(., class) %>% print()}

##      wind_gust_dir wind_dir_9am wind_dir_3pm
## [1,] "ordered"     "ordered"    "ordered"
## [2,] "factor"      "factor"     "factor"
```

Again we obtain a distribution of the variables to confirm that all we have changed is the data type.

```
# Verify the distribution has not changed.

ds %>%
  select(contains("_dir")) %>%
  sapply(table)

##     wind_gust_dir wind_dir_9am wind_dir_3pm
## N            8922        11339         8520
## NNE          6260         7763         6332
## NE           6822         7303         7938
## ENE          7669         7436         7456
## E            8628         8622         8024
....
```

There are two other variables that have been identified as character data types: `evaporation` and `sunshine`. If we look at the dataset we see they have missing values.

```
# Note the remaining character variables to be dealt with.

cvars <- c("evaporation", "sunshine")
```

```
# Review their values.

head(ds[cvars])

## # A tibble: 6 x 2
##   evaporation sunshine
##         <chr>    <chr>
## 1        <NA>     <NA>
## 2        <NA>     <NA>
## 3        <NA>     <NA>
## 4        <NA>     <NA>
## 5        <NA>     <NA>
## 6        <NA>     <NA>

sample_n(ds[cvars], 6)

## # A tibble: 6 x 2
##   evaporation sunshine
##         <chr>    <chr>
## 1        <NA>     <NA>
## 2         5.8       12
## 3           2        9
## 4         1.8      8.4
## 5         2.2        0
## 6         4.8        0
```

The heuristic used to determine the data type when `readr::read_csv()` ingests the data only looks at a subset of all the data before it determines the data types. In this case they were missing for the early observations and so in the absence of further information they were represented as character. We need to convert them to numeric.

```
# Check the current class of the variables.

ds[cvars] %>% sapply(class)

## evaporation    sunshine
## "character" "character"

# Convert to numeric.
```

```
ds[cvars] %<>% sapply(as.numeric)

# Confirm the conversion.

ds[cvars] %>% sapply(class)

## evaporation    sunshine
##   "numeric"   "numeric"
```

We have now dealt with all of the `character` variables converting them to `factors` or `numerics` on a case-by-case basis on our understanding of the data.

Normalise Factors

Some variables will have levels with spaces, and mixture of cases, etc. We may like to normalise the levels for each of the categoric variables. For very large datasets this can take some time and so we may choose to be selective if there are many factors.

```
# Note which variables are categoric.

ds %>%
  sapply(is.factor) %>%
  which() %T>%
  print() ->
catc

##       location wind_gust_dir  wind_dir_9am  wind_dir_3pm
##              2             8            10            11
##     rain_today rain_tomorrow
##             22            24

# Normalise the levels of all categoric variables.

for (v in catc)
  levels(ds[[v]]) %<>% normVarNames()
```

To confirm we can review the categoric variables.

```
glimpse(ds[catc])
```

```
## Observations: 138,307
## Variables: 6
## $ location      <fctr> albury, albury, albury, albury, alb...
## $ wind_gust_dir <ord> w, wnw, wsw, ne, w, wnw, w, w, nnw, ...
## $ wind_dir_9am  <ord> w, nnw, w, se, ene, w, sw, sse, se, ...
## $ wind_dir_3pm  <ord> wnw, wsw, wsw, e, nw, w, w, w, nw, s...
....
```

Ensure Target is a Factor

Many data mining tasks can be expressed as building classification models. For such models we want to ensure the target is categoric. Often it is 0/1 and hence is loaded as numeric. In such cases we could tell our model algorithm to explicitly do classification or else set the target using `base::as.factor()` in the formula. Nonetheless it is generally cleaner to do this here and note that this code has no effect if the target is already categoric.

```
# Note the target variable.

target <- "rain_tomorrow"

# Ensure the target is categoric.

ds[[target]] %<>% as.factor()

# Confirm the distribution.

ds[target] %>% table()

## .
##     no    yes
## 104493  30283
```

It is always a good idea to visualise the distribution of the target (and other) variables using ggplot2. We can pipe the dataset into `ggplot2::ggplot()` whereby the target is associated through `ggplot2::aes_string()` (the aesthetics) with the x-axis of the plot. To this we add a graphics layer using `ggplot2::geom_bar()` to produce the bar chart, with bars having `width=` `0.2` and a `fill=` color of `"grey"`. The resulting plot can be seen in Figure 3.1. With

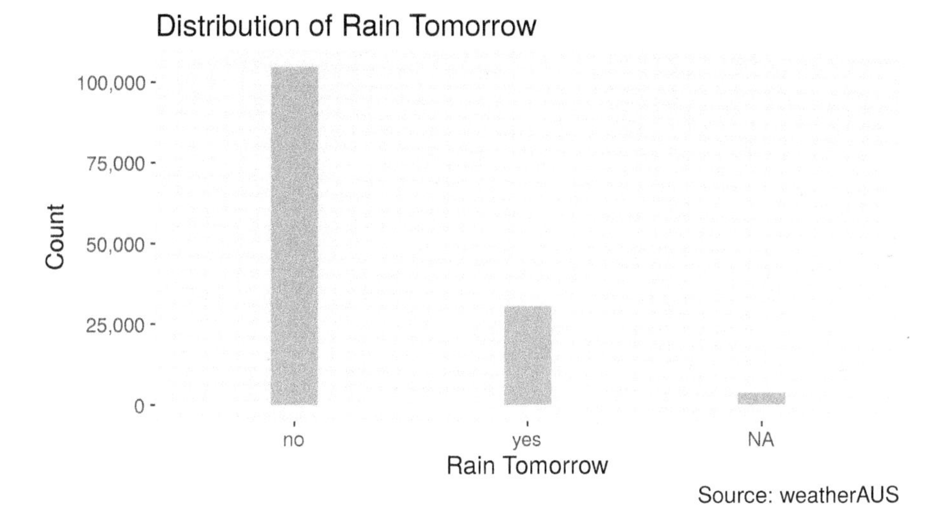

Figure 3.1: *Target variable distribution. Plotting the distribution is useful to gain an insight into the number of observations in each category. As is the case here we often see a skewed distribution.*

some surprise we note that there are missing values in the dataset. We will deal with the missing values (`NA`) shortly.

```
ds %>%
  ggplot(aes_string(x=target)) +
  geom_bar(width=0.2, fill="grey") +
  theme(text=element_text(size=14)) +
  scale_y_continuous(labels=comma) +
  labs(title   = "Distribution of Rain Tomorrow",
       x       = "Rain Tomorrow",
       y       = "Count",
       caption = "Source: weatherAUS")
```

3.4 Variable Roles

Now that we have a basic idea of the size and shape and contents of the dataset and have performed some basic data type identification and cleaning, we are in a position to identify the roles played by the variables within the dataset. We record the list of available variables so that we might reference them shortly.

```
# Note the available variables.

ds %>%
  names() %T>%
  print() ->
vars

##  [1] "date"            "location"        "min_temp"
##  [4] "max_temp"        "rainfall"        "evaporation"
##  [7] "sunshine"        "wind_gust_dir"   "wind_gust_speed"
## [10] "wind_dir_9am"    "wind_dir_3pm"    "wind_speed_9am"
## [13] "wind_speed_3pm"  "humidity_9am"    "humidity_3pm"
## [16] "pressure_9am"    "pressure_3pm"    "cloud_9am"
## [19] "cloud_3pm"       "temp_9am"        "temp_3pm"
## [22] "rain_today"      "risk_mm"         "rain_tomorrow"
```

By this stage of the project we will usually have identified a business problem that is the focus of attention. In our case we will assume it is to build a predictive analytics model to predict the chance of it raining tomorrow given the observation of today's weather. In this case, the variable `rain_tomorrow` is the *target variable*. Given today's observations of the weather this is what we want to predict. The dataset we have is then a *training dataset* of historic observations. The task in model building is to identify any patterns among the other observed variables that suggest that it rains the following day.

We also take the opportunity here to move the target variable to be the first in the vector of variables recorded in `vars`. This is common practice where the first variable in a dataset is the target and the remainder are the variables that will be used to build a model. Another common practise is for the target to be the final column of the dataset.

```
# Place the target variable at the beginning of the vars.

c(target, vars) %>%
  unique() %T>%
  print() ->
```

```
vars

##  [1] "rain_tomorrow"    "date"             "location"
##  [4] "min_temp"         "max_temp"         "rainfall"
##  [7] "evaporation"      "sunshine"         "wind_gust_dir"
## [10] "wind_gust_speed" "wind_dir_9am"     "wind_dir_3pm"
## [13] "wind_speed_9am"  "wind_speed_3pm"   "humidity_9am"
## [16] "humidity_3pm"    "pressure_9am"     "pressure_3pm"
....
```

Notice the use of `base::unique()` simply to remove the original occurrence of the target variable.

Another variable that we observe as relating to the outcome rather than to today's observations is `risk_mm`. From the business context we would learn that this records the amount of rain that fell "tomorrow". We refer to this as a *risk variable*. It is a measure of the impact or risk of the outcome we are predicting (whether it rains tomorrow). The risk is an output variable and should not be used as an input to the modelling—it is not an independent variable. In other circumstances it might actually be treated as the target variable.

```
# Note the risk variable - measures the severity of the outcome.

risk <- "risk_mm"
```

Finally from our observations so far we note that the variable `date` acts as an identifier as does the variable `location`. Given a `date` and a `location` we have an observation of the remaining variables. We note these two variables as identifiers. Identifiers would not usually be used as independent variables for building predictive analytics models.

```
# Note any identifiers.

id <- c("date", "location")
```

3.5 Feature Selection

We now move on to identifying variables (features or columns of the dataset) that are irrelevant or inappropriate for modelling.

IDs and Outputs

We start by noting that we should ignore all identifiers and any risk variable (which will be an output variable rather than an input variable). These variables should be ignored in our modelling. Always watch out for treating output variables as inputs to modelling—this is a surprisingly common trap for beginners.

We will build a vector of the names of the variables to ignore. Above we have already recorded the `id` variables and (optionally) the `risk`. Here we join them into a new vector using `dplyr::union()` which performs a set union operation—that is, it joins the two arguments together and removes any repeated variables.

```
# Initialise ignored variables: identifiers and risk.

id %>%
  union(if (exists("risk")) risk) %T>%
  print() ->
ignore

## [1] "date"     "location" "risk_mm"
```

We might also check for variables that have a unique value for every observation. These are often identifiers and if so they are candidates for ignoring.

We begin in the following code block by defining a helper function that given a vector of data it will return the number of unique values found in that vector. This helper function is then deployed in the following pipeline to identify those `vars` which have as many unique values as there are rows in the dataset. The pipeline is explained next. This is our first example of defining our own function.

```
# Helper function to count the number of distinct values.

count_unique <- function(x)
{
  x %>% unique() %>% length()
}

# Heuristic for candidate indentifiers to possibly ignore.

ds[vars] %>%
  sapply(count_unique) %>%
  equals(nrow(ds)) %>%
  which() %>%
  names() %T>%
  print() ->
ids

## character(0)

# Add them to the variables to be ignored for modelling.

ignore <- union(ignore, ids) %T>% print()

## [1] "date"     "location" "risk_mm"
```

We can step through this code line by line to understand its workings. To find the candidate identifiers we retain just the `vars` from the dataset and pipe this subset of the original dataset `ds` through to `base::sapply()`. The function `base::sapply()` applies a supplied function to every column of the provided dataset. The function we supply is the helper function we defined. The (x) represents a single column of the dataset at a time. Thus, for each column we identify the `base::unique()` values in that column and then return the `base::length()` of the vector of unique values.

The `base::sapply()` sends the vector of lengths (the number of unique values for each of the columns) on to the next operation through the pipeline. The follow-on operation tests if the calculated number of unique values `magrittr::equals()` the number of rows in the dataset as calculated using `base::nrow()`. The resulting vector of logical values is then piped to `base::which()` to retain those that are TRUE—those that have as many unique values as there are rows in the dataset.

Finally, we extract the base::names() of these variables and store them as the variable ids after printing them for information purposes.

We have strung together a series of operations here with each operation piping data on to the next operation. It is worth taking a little time to understand the sequence as a single sentence in the grammar of ***data wrangling***. As we interact with our data we typically build the sequence adding one extra process at a time in the R Console, confirming the results as we go.

The pipeline has identified no variables as potential identifiers themselves in this dataset; hence, the character(0) result. Below we choose observations from a single location and process that data through the above pipeline to illustrate the selection of date as having a unique value for every observation.

```
# Engineer the data to illustrate identifier selection.

ods <- ds # Take a backup copy of the original dataset.

ds %<>% filter(location=="sydney")

ds[vars] %>%
  sapply(count_unique) %>%
  equals(nrow(ds)) %>%
  which() %>%
  names()

## [1] "date"

ds <- ods # Restore the original dataset.
```

All Missing

We next remove any variable where all of the values are missing. Our pipeline here counts the number of missing values for each variable and then lists the names of those variables that have no values. We introduce another small helper function to count the number of missing values for a vector.

```
# Helper function to count the number of values missing.

count_na <- function(x)
{
  x %>% is.na() %>% sum()
}

# Identify variables with only missing values.

ds[vars] %>%
  sapply(count_na) %>%
  equals(nrow(ds)) %>%
  which() %>%
  names() %T>%
  print() ->
missing

## character(0)

# Add them to the variables to be ignored for modelling.

ignore %<>% union(missing) %T>% print()

## [1] "date"     "location" "risk_mm"
```

Again there are no variables that are completely missing in the **weatherAUS** dataset but in general it is worth checking. Below we engineer a dataset with all missing values for some variables to illustrate the pipeline in action.

```
# Engineer the dataset to illustrate missing columns.

ods <- ds # Take a backup copy of the dataset.

ds %<>% filter(location=="albury")

ds[vars] %>%
  sapply(count_na) %>%
  equals(nrow(ds)) %>%
  which() %>%
  names()

## [1] "evaporation" "sunshine"

ds <- ods # Restore the dataset.
```

Note that it is also adding to our knowledge of this dataset that there are locations for which some variables are not observed or recorded. This may play a role in understanding how to model the data.

Many Missing

It is also useful to identify those variables which are very sparse—that have mostly missing values. We can decide on a threshold of the proportion missing above which to ignore the variable as not likely to add much value to our analysis. For example, we may want to ignore variables with more than 80% of the values missing:

```
# Identify a threshold above which proportion missing is fatal.

missing.threshold <- 0.8

# Identify variables that are mostly missing.

ds[vars] %>%
  sapply(count_na) %>%
  '>'(missing.threshold*nrow(ds)) %>%
  which() %>%
  names() %T>%
  print() ->
mostly

## character(0)

# Add them to the variables to be ignored for modelling.

ignore <- union(ignore, mostly) %T>% print()

## [1] "date"     "location" "risk_mm"
```

Here again we identify no variables that have a high proportion of missing observations.

Too Many Levels

Another issue we often come across in our datasets are factors that have very many levels. We might want to ignore such variables (or perhaps group them appropriately). Here we simply identify them and add them to the list of variables to ignore:

```
# Helper function to count the number of levels.

count_levels <- function(x)
{
  ds %>% extract2(x) %>% levels() %>% length()
}

# Identify a threshold above which we have too many levels.

levels.threshold <- 20

# Identify variables that have too many levels.

ds[vars] %>%
  sapply(is.factor) %>%
  which() %>%
  names() %>%
  sapply(count_levels) %>%
  '>='(levels.threshold) %>%
  which() %>%
  names() %T>%
  print() ->
too.many

## [1] "location"

# Add them to the variables to be ignored for modelling.

ignore <- union(ignore, too.many) %T>% print()

## [1] "date"     "location" "risk_mm"
```

The variable `location` is identified as having too many levels and is thus added to the ignore list though since it is already on that list there is no change to it.

Constants

We should also ignore variables with constant values as they generally add no extra information to the analysis.

```
# Helper function to test if all values in vector are the same.

all_same <- function(x)
{
  all(x == x[1L])
}

# Identify variables that have a single value.

ds[vars] %>%
  sapply(all_same) %>%
  which() %>%
  names() %T>%
  print() ->
constants

## character(0)

# Add them to the variables to be ignored for modelling.

ignore <- union(ignore, constants) %T>% print()

## [1] "date"     "location" "risk_mm"
```

There are no constants found in this dataset.

Correlated Variables

It is often useful to reduce the number of variables we are modelling by identifying and removing highly correlated variables. Such variables will often record the same information but in different ways. Correlated variables can often arise when we combine data from different sources.

First we will identify the numeric variables on which we will calculate correlations. We start by removing the ignored variables from the dataset. We then identify the numeric variables by `base::sapply()`ing the function `base::is.numeric()` to the dataset then find `base::which()` variables are numeric. The variable names are stored into the variable `numc`.

```
# Note which variables are numeric.

vars %>%
  setdiff(ignore) %>%
  magrittr::extract(ds, .) %>%
  sapply(is.numeric) %>%
  which() %>%
  names() %T>%
  print() ->
numc

##  [1] "min_temp"        "max_temp"        "rainfall"
##  [4] "evaporation"     "sunshine"        "wind_gust_speed"
##  [7] "wind_speed_9am"  "wind_speed_3pm"  "humidity_9am"
## [10] "humidity_3pm"    "pressure_9am"    "pressure_3pm"
## [13] "cloud_9am"       "cloud_3pm"       "temp_9am"
## [16] "temp_3pm"
```

We can then calculate the correlation between the numeric variables by selecting the numeric columns from the dataset and passing that through to `stats::cor()`. This generates a matrix of pairwise correlations based on only the complete observations so that observations with missing values are ignored.

We set the upper triangle of the correlation matrix to `NA`'s as they are a mirror of the values in the lower triangle and thus redundant. Notice that with `diag=TRUE` this includes the diagonals of the matrix being set to `NA` as they will always be perfect correlations (1).

Next we ensure the values are positive using `base::abs()`. We also ensure we have a `base::data.frame()` which we convert to a `dplyr::tbl_df()`. The dataset column names need to be reset appropriately using `magrittr::set_colnames()`. We `dplyr::mutate()` the dataset by adding a new column, then `tidyr::gather()` the dataset. Missing correlations are omitted using `stats::na.omit()`. Finally, the rows are `dplyr::arrange()`'d with the highest absolute correlations appearing first.

```
# For numeric variables generate a table of correlations
```

```
ds[numc] %>%
  cor(use="complete.obs") %>%
  ifelse(upper.tri(., diag=TRUE), NA, .) %>%
  abs() %>%
  data.frame() %>%
  tbl_df() %>%
  set_colnames(numc) %>%
  mutate(var1=numc) %>%
  gather(var2, cor, -var1) %>%
  na.omit() %>%
  arrange(-abs(cor)) %T>%
  print() ->
mc

## # A tibble: 120 x 3
##             var1           var2       cor
##            <chr>          <chr>     <dbl>
## 1       temp_3pm       max_temp 0.9851472
## 2   pressure_3pm   pressure_9am 0.9620731
## 3       temp_9am       min_temp 0.9069890
## 4       temp_9am       max_temp 0.8936331
## 5       temp_3pm       temp_9am 0.8710598
## 6       max_temp       min_temp 0.7497200
## 7       temp_3pm       min_temp 0.7271346
....
```

This is quite a complex pipeline. It is worth taking time to understand the sequence as a single sentence in the grammar of data wrangling. Importantly it should be noted that we build up such a command sequence interactively, adding one new command in the pipeline sequence at a time until we have our final desired outcome. It is strongly recommended that you replicate building the sequence one step at a time to review and understand the result after each step.

From the final result we can identify pairs of variables where we might want to keep one but not the other variable because they are highly correlated. We will select them manually since it is a judgement call. Normally we might limit the removals to those correlations that are 0.90 or more. We should confirm that the three most highly correlated variables here make intuitive sense.

```
# Note the correlated variables that are redundant.

correlated <- c("temp_3pm", "pressure_3pm", "temp_9am")

# Add them to the variables to be ignored for modelling.

ignore <- union(ignore, correlated) %T>% print()

## [1] "date"          "location"      "risk_mm"        "temp_3pm...
## [5] "pressure_3pm" "temp_9am"
```

Removing Variables

Once we have identified all of the variables to ignore we remove them from our list of variables to use.

```
# Check the number of variables currently.

length(vars)

## [1] 24

# Remove the variables to ignore.

vars %<>% setdiff(ignore) %T>% print()

##  [1] "rain_tomorrow"   "min_temp"        "max_temp"
##  [4] "rainfall"        "evaporation"     "sunshine"
##  [7] "wind_gust_dir"   "wind_gust_speed" "wind_dir_9am"
## [10] "wind_dir_3pm"    "wind_speed_9am"  "wind_speed_3pm"
## [13] "humidity_9am"    "humidity_3pm"    "pressure_9am"
## [16] "cloud_9am"       "cloud_3pm"       "rain_today"

# Confirm they are now ignored.

length(vars)

## [1] 18
```

Algorithmic Feature Selection

There are many R packages available to support the preparation of our datasets and over time you will find packages that suit your needs. As you do so they can be added to your version of the template for data wrangling. For example, a useful package is FSelector which provides functions to identify subsets of variables that might be most effective for modelling. We can use this (and other packages) to further assist us in reducing the variables for modelling.

As an example we can use FSelector::**cfs()** to identify a subset of variables to consider for use in modelling by using correlation and entropy.

```
# Construct the formulation of the modelling to undertake.

form <- formula(target %s+% " ~ .") %T>% print()

## rain_tomorrow ~ .

# Use correlation search to identify key variables.

cfs(form, ds[vars])

## [1] "sunshine"     "humidity_3pm" "rain_today"
```

Notice the use of the stringi::**%s+%** operator as a convenience to concatenate strings together to produce a formula that indicates we will model the target variable using all of the other variables of the dataset.

A second example lists the variable importance using FSelector::**information.gain()** to advise a useful subset of variables.

```
# Use information gain to identify variable importance.

information.gain(form, ds[vars]) %>%
  rownames_to_column("variable") %>%
  arrange(-attr_importance)

##          variable attr_importance
```

```
## 1      humidity_3pm   0.113700324
## 2          sunshine   0.058452293
## 3          rainfall   0.056252382
## 4         cloud_3pm   0.053338637
## 5        rain_today   0.044860122
## 6      humidity_9am   0.039193871
## 7         cloud_9am   0.036895020
## 8      pressure_9am   0.028837842
## 9   wind_gust_speed   0.025299064
## 10         max_temp   0.014812106
## 11     wind_dir_9am   0.008286797
## 12      evaporation   0.006114435
## 13    wind_gust_dir   0.005849817
## 14         min_temp   0.005012095
## 15   wind_speed_3pm   0.004941750
## 16     wind_dir_3pm   0.004717975
## 17   wind_speed_9am   0.003778012
```

The two measures are somewhat consistent in this case. The variables identified by `FSelector::cfs()` are mostly the more important variables identified by `FSelector::information.gain()`. However `rain_today` is a little further down on the information gain list.

3.6 Missing Data

A common task is to deal with missing values. Here we remove observations with a missing target. As with any missing data we should also analyse whether there is any pattern to the missing targets. This may be indicative of a systematic data issue rather than simple randomness. It is important to investigate further why the data is systematically missing. Often this will also lead to a better understanding of the data and how it was collected.

```
# Check the dimensions to start with.

dim(ds)

## [1] 138307     24
```

```
# Identify observations with a missing target.

missing_target <- ds %>% extract2(target) %>% is.na()

# Check how many are found.

sum(missing_target)

## [1] 3531

# Remove observations with a missing target.

ds %<>% filter(!missing_target)

# Confirm the filter delivered the expected dataset.

dim(ds)

## [1] 134776     24
```

Missing values in the input variables are also an issue for some but not all algorithms. For example, the traditional ensemble model building algorithm randomForest::**randomForest()** omits observations with missing values by default whilst rpart::**rpart()** has a particularly well-developed approach to dealing with missing values.

In the previous section we removed variables with many missing values noting that this is not always appropriate. We may want to instead impute missing values in the data (noting also that it is not always wise to do so as it is effectively inventing data).

Here we illustrate a process for imputing missing values using randomForest::**na.roughfix()**. As the name suggests, this function provides a basic algorithm for imputing missing values. We will demonstrate the process but then restore the original dataset rather than have the imputations included in our actual dataset.

```
# Backup the dataset so we can restore it as required.

ods <- ds

# Count the number of missing values.
```

```
ds[vars] %>%  is.na() %>% sum() %>% comcat()

## 278,293

# Impute missing values.

ds[vars] %<>% na.roughfix()

# Confirm that no missing values remain.

ds[vars] %>%  is.na() %>% sum() %>% comcat()

## 0

# Restore the original dataset.

ds <- ods
```

An alternative might be to remove observations that have missing values. We use `stats::na.omit()` to identify the rows to omit based on the `vars` to be included for modelling. The list of rows to omit is stored as the `na.action` attribute of the returned object. We can then remove these observations from the dataset. We start again by keeping a copy of the original dataset to restore later. We also initialise a list of row indicies that we will (omit) from the dataset.

```
# Backup the dataset so we can restore it as required.

ods <- ds

# Initialise the list of observations to be removed.

omit <- NULL

# Review the current dataset.

ds[vars] %>% nrow()

## [1] 134776

ds[vars] %>% is.na() %>% sum() %>% comcat()
```

```
## 278,293

# Identify any observations with missing values.

mo <- attr(na.omit(ds[vars]), "na.action")

# Record the observations to omit.

omit <- union(omit, mo)

# If there are observations to omit then remove them.

if (length(omit)) ds <- ds[-omit,]

# Confirm the observations have been removed.

ds[vars] %>% nrow() %>% comcat()

## 54,757

ds[vars] %>% is.na() %>% sum()

## [1] 0

# Restore the original dataset.

ds   <- ods
omit <- NULL
```

By restoring the dataset to its original contents to continue with our analysis we are deciding not to omit any observations at this time.

3.7 Feature Creation

Another important task for the data scientist is to create new features from the provided data where appropriate. We will illustrate with two examples. Note that we will often iterate over the feature creation process many times during the life cycle of a project. New derived features will become identified as we gain insights into the data and through our modelling.

Derived Features

From a review of the data we note that each observation has a date associated with it. Unless we are specifically performing time series analysis (and indeed this would be an appropriate analysis to consider for this dataset) some derived features may be useful for our model building rather than using the specific dates as input variables.

Here we add two derived features to our dataset: year and season. The decision to add these was made after we initially began exploring the data and building our initial predictive models. Feedback from our domain expert suggested that the changing pattern over the years is of interest and that predicting rain will often involve seasonal adjustments.

The dataset is `dplyr::mutate()`'d to add these two features. The `year` is a simple extraction from the date using `base::format()`. To compute the season we extract the month `base::as.integer()` which is then an index to `base::switch()` to a specific season depending on the month.

```
ds %<>%
  mutate(year   = factor(format(date, "%Y")),
         season = format(ds$date, "%m") %>%
                  as.integer() %>%
                  sapply(function(x)
                    switch(x,
                           "summer", "summer", "autumn",
                           "autumn","autumn", "winter",
                           "winter", "winter", "spring",
                           "spring", "spring", "summer")) %>%
                  as.factor()) %T>%
  {select(., date, year, season) %>% sample_n(10) %>% print()}
```

```
## # A tibble: 10 x 3
##          date   year season
##        <date> <fctr> <fctr>
##  1 2014-11-13   2014 spring
##  2 2014-10-15   2014 spring
```

```
## 3 2010-12-18  2010 summer
## 4 2010-02-06  2010 summer
## 5 2016-01-07  2016 summer
## 6 2015-01-01  2015 summer
## 7 2009-11-22  2009 spring
## 8 2013-09-28  2013 spring
## 9 2009-06-09  2009 winter
## 10 2012-05-27  2012 autumn
```

The final line of code prints a random sample of the original and new features. It is critical to confirm the results are as expected—a sanity check.

The introduced variables will have different roles which are recorded appropriately.

```
vars %<>% c("season")
id   %<>% c("year")
```

Model-Generated Features

In addition to developing features derived from other features we will sometimes find it useful to include model-generated features. A common one is to cluster observations or some aggregation of observations into similar groups. We then uniquely identify each group (for example, by numbering each group) and add this grouping as another column or variable within our dataset.

A cluster analysis (also called segmentation) provides a simple mechanism to identify groups and allows us to then visualise within those groups, for example, rather than trying to visualise the whole dataset at one time. We will illustrate the process of a cluster analysis over the numeric variables within our dataset. The aim of the cluster analysis is to group the locations so that within a group the locations will have similar values for the numeric variables and between the groups the numeric variables will be more dissimilar. The traditional clustering algorithm is `stats::kmeans()`.

In the following code block we specify the number of clusters that we wish to create, storing that as the variable `NCLUST`. We then select the `numc` (numeric) variables from the dataset `ds`, `dplyr::group_by()` the `location` and then `dplyr::summarise_all()` variables. As an interim step we store for later use the names of the locations. We then continue by removing `location` from the dataset leaving just the numeric variables. The variables are then rescaled so that all values across the different variables are in the same range. This rescaling is required for cluster analysis so as not to introduce bias due to very differently scaled variables. The `stats::kmeans()` cluster analysis is then performed. We `base::print()` the results and `magrittr::extract2()` just the cluster number for each location.

```
# Reset the random number generator seed for repeatablity.

set.seed(7465)

# Cluster the numeric data per location.

NCLUST <- 5

ds[c("location", numc)] %>%
  group_by(location) %>%
  summarise_all(funs(mean(., na.rm=TRUE))) %T>%
  {locations <<- .$location} %>% # Store locations for later.
  select(-location) %>%
  sapply(function(x) ifelse(is.nan(x), 0, x)) %>%
  as.data.frame() %>%
  sapply(scale) %>%
  kmeans(NCLUST)  %T>%
  print() %>%
  extract2("cluster")->
cluster

## K-means clustering with 5 clusters of sizes 4, 22, 10, 8, 5
##
## Cluster means:
##     min_temp   max_temp     rainfall evaporation    sunshine
## 1 -0.6271436 -0.5345409  0.061972675  -1.2699891 -1.21861982
## 2 -0.3411683 -0.5272989 -0.007762188   0.1137179  0.09919753
....

head(cluster)

## [1] 3 2 2 3 4 2
```

The cluster numbers are now associated with each location as stored within the vector using their base::**names()**. We can then dplyr::**mutate()** the dataset by adding the new cluster number indexed by the location for each observation.

```
# Index the cluster vector by the appropriate locations.

names(cluster) <- locations

# Add the cluster to the dataset.

ds %<>% mutate(cluster="area" %>%
                 paste0(cluster[ds$location]) %>%
                 as.factor)

# Check clusters.

ds %>% select(location, cluster) %>% sample_n(10)

## # A tibble: 10 x 2
##          location cluster
##            <fctr>  <fctr>
##  1       richmond   area2
##  2          cobar   area3
##  3  mount_gambier   area2
##  4           nhil   area4
##  5         sydney   area2
##  6     launceston   area2
##  7   mount_ginini   area1
##  8     wollongong   area2
##  9     gold_coast   area4
## 10         hobart   area2
```

The introduced variable's role in modelling is recorded appropriately.

```
vars %<>% c("cluster")
```

A quick sanity check will indicate that basically the clustering looks okay, grouping locations that the domain expert agrees are similar in terms of weather patterns.

```
# Check that the clustering looks okay.

cluster[levels(ds$location)] %>% sort()

##      mount_ginini          newcastle            penrith
##                 1                  1                  1
##       salmon_gums             albany             albury
##                 1                  2                  2
##          ballarat            bendigo           canberra
##                 2                  2                  2
....
```

3.8 Preparing the Metadata

Metadata is data about the data. We now record data about our dataset that we use later in further processing and analysis.

Variable Types

We identify the variables that will be used to build analytic models that provide different kinds of insight into our data. Above we identified the variable roles such as the target, a risk variable and the ignored variables. From an analytic modelling perspective we also identify variables that are the model inputs (also called the independent variables). We record then both as a vector of characters (the variable names) and a vector of integers (the variable indicies).

```
vars %>%
  setdiff(target) %T>%
  print() ->
inputs

##  [1] "min_temp"         "max_temp"         "rainfall"
##  [4] "evaporation"      "sunshine"         "wind_gust_dir"
##  [7] "wind_gust_speed" "wind_dir_9am"     "wind_dir_3pm"
## [10] "wind_speed_9am"  "wind_speed_3pm"   "humidity_9am"
## [13] "humidity_3pm"     "pressure_9am"     "cloud_9am"
```

```
## [16] "cloud_3pm"        "rain_today"       "season"
## [19] "cluster"
```

The integer indices are determined from the base::**names()** of the variables in the original dataset. Note the use of USE.NAMES= from base::**sapply()** to turn off the inclusion of names in the resulting vector to keep the result as a simple vector.

```
inputs %>%
  sapply(function(x) which(x == names(ds)), USE.NAMES=FALSE)%T>%
  print() ->
inputi

##  [1]  3  4  5  6  7  8  9 10 11 12 13 14 15 16 18 19 22 26 27
```

For convenience we also record the number of observations:

```
ds %>%
  nrow() %T>%
  comcat() ->
nobs

## 134,776
```

Next we report on the dimensions of various data subsets primarily to confirm that the dataset appears as we expect:

```
# Confirm various subset sizes.

dim(ds)            %>% comcat()

## 134,776 27

dim(ds[vars])      %>% comcat()

## 134,776 20

dim(ds[inputs]) %>% comcat()

## 134,776 19

dim(ds[inputi]) %>% comcat()

## 134,776 19
```

Numeric and Categoric Variables

Sometimes we need to identify the numeric and categoric variables for separate handling. Many cluster analysis algorithms, for example, only deal with numeric variables. Here we identify them both by name (a character string) and by index. Note that when using the index we have to assume the variables remain in the same order within the dataset and all variables are present. Otherwise the indicies will get out of sync.

```
# Identify the numeric variables by index.

ds %>%
  sapply(is.numeric) %>%
  which() %>%
  intersect(inputi) %T>%
  print() ->
numi

## [1]  3  4  5  6  7  9 12 13 14 15 16 18 19

# Identify the numeric variables by name.

ds %>%
  names() %>%
  extract(numi) %T>%
  print() ->
numc

##  [1] "min_temp"        "max_temp"        "rainfall"
##  [4] "evaporation"     "sunshine"        "wind_gust_speed"
##  [7] "wind_speed_9am"  "wind_speed_3pm"  "humidity_9am"
## [10] "humidity_3pm"    "pressure_9am"    "cloud_9am"
## [13] "cloud_3pm"

# Identify the categoric variables by index.

ds %>%
  sapply(is.factor) %>%
  which() %>%
  intersect(inputi) %T>%
  print() ->
cati
```

```
## [1]  8 10 11 22 26 27

# Identify the categoric variables by name.

ds %>%
  names() %>%
  extract(cati) %T>%
  print() ->
catc

## [1] "wind_gust_dir" "wind_dir_9am"  "wind_dir_3pm"
## [4] "rain_today"    "season"        "cluster"
```

3.9 Preparing for Model Building

Our end goal is to build a model based on the data we have prepared. A ***model*** will capture knowledge about the world that the data represents. The final two tasks in preparing our data for modelling are to specify the form of the model to be built and to identify the actual observations from which the model is to be built. For the latter task we will partition the dataset into three subsets. Whilst this is primarily useful for model building, it may sometimes be useful to explore a random subset of the whole dataset so that our interactions are more interactive, particularly when dealing with large datasets.

Formula to Describe the Model

A ***formula*** is used to identify what it is that we will model from the supplied data. Typically we identify a ***target variable*** which we will model based on other ***input variables***. We can construct a `stats::formula()` automatically from a dataset if the first column of the dataset is the target variable and the remaining columns are the input variables. A simple selection of the columns in this order will generate the initial formula automatically. Earlier we engineered the variable `vars` to be in the required order.

```
ds[vars] %>%
  formula() %>%
  print() ->
form

## rain_tomorrow ~ min_temp + max_temp + rainfall + evaporati...
##     sunshine + wind_gust_dir + wind_gust_speed + wind_dir_...
##     wind_dir_3pm + wind_speed_9am + wind_speed_3pm + humid...
##     humidity_3pm + pressure_9am + cloud_9am + cloud_3pm + ...
##     season + cluster
## <environment: 0x5640bab65ef8>
```

The notation used to express the formula begins with the name of a target (`rain_tomorrow`) followed by a tilde (`~`) followed by the variables that will be used to model the target, each separated by a plus (`+`). The formula here indicates that we aim to build a model that captures the knowledge required to predict the outcome `rain_tomorrow` from the provided input variables (details of today's weather). This kind of model is called a classification model and can be compared to regression models, for example, which predict numeric outcomes. Specifically, with just two values to be predicted, we call this binary classification. It is generally the simplest kind of modelling task but also a very common task.

Training, Validation and Testing Datasets

Models are built using a machine learning algorithm which learns from the dataset of historic observations. A common methodology for building models is to partition the available data into a ***training dataset*** and a ***testing dataset***. We will build (or ***train***) a model using the training dataset and then test how good the model is by applying it to the testing dataset. Typically we also introduce a third dataset called the ***validation dataset***. This is used during the building of the model to assist in tuning the algorithm through trialling different parameters to the machine learning algorithm. In this way we search for the best model using the validation dataset and then obtain a measure of the performance of the final model using the testing dataset.

The original dataset is partitioned randomly into the three subsets. To ensure we can repeatably reproduce our results we will first initiate a random number sequence with a randomly selected seed. In this way we can replicate the examples presented in this book by ensuring the same random subset is selected each time. We will initialise the random number generator with a specific seed using `base::set.seed()`. For no particular reason we choose 42.

```
# Initialise random numbers for repeatable results.

seed <- 42
set.seed(seed)
```

We are now ready to partition the dataset into the two or three subsets. The first is typically a 70% random sample for building the model (the training dataset). The second and third consist of the remainder, used to tune and then estimate the expected performance of the model (the validation and testing datasets).

Rather than actually creating three subsets of the dataset, we simply record the index of the observations that belong to each of the three subsets.

```
# Partition the full dataset into three.

nobs %>%
  sample(0.70*nobs) %T>%
  {length(.) %>% comcat()} %T>%
  {sort(.) %>% head(30) %>% print()} ->
train

## 94,343
##  [1]  1  4  7  9 10 11 12 13 14 15 16 17 18 19 21 22 23 24 26
## [20] 28 31 32 33 35 37 40 41 42 43 45

nobs %>%
  seq_len() %>%
  setdiff(train) %>%
  sample(0.15*nobs) %T>%
  {length(.) %>% comcat()} %T>%
  {sort(.) %>% head(15) %>% print()} ->
validate

## 20,216
##  [1]  2  3  6 20 25 27 29 34 39 44 49 65 66 71 81
```

```
nobs %>%
  seq_len() %>%
  setdiff(union(train, validate)) %T>%
  {length(.) %>% comcat()} %T>%
  {head(.) %>% print(15)} ->
test
```

```
## 20,217
## [1]  5  8 30 36 38 48
```

We take a moment here to record the actual target values across the three datasets as these will be used in the evaluations performed in later chapters. A secondary target variable is noted, referred to as the risk variable. This is a measure of the impact or size of the outcome and is common in insurance, financial risk and fraud analysis. For our sample dataset, the risk variable is `risk_mm` which reports the amount of rain recorded tomorrow.

Notice the correspondence between risk values (the amount of rain) and the target, where 0.0 mm of rain corresponds to the target value `no`. In particular though also note that small amounts of rain (e.g., 0.2 mm and 0.4 mm) are treated as no rain.

```
# Cache the various actual values for target and risk.

tr_target <- ds[train,][[target]] %T>%
             {head(., 20) %>% print()}
```

```
##  [1] no  no  no  no  no  no  no  no  no  no  no  no  no  no
## [15] no  no  no  no  yes no
....
```

```
tr_risk   <- ds[train,][[risk]] %T>%
             {head(., 20) %>% print()}
```

```
##  [1] 0.0 0.0 0.0 0.0 0.4 0.0 0.2 0.0 0.0 0.0 0.0 0.2 0.2 1.0
## [15] 0.2 0.0 0.0 0.0 1.2 0.0
....
```

```
va_target <- ds[validate,][[target]] %T>%
             {head(., 20) %>% print()}
```

```
##  [1] yes no  no  no  no  no  no  no  no  no  yes no  no  no
## [15] no  yes no  no  no  yes
....

va_risk   <- ds[validate,][[risk]] %T>%
             {head(., 20) %>% print()}

##  [1] 9.8 0.0 0.0 0.0 0.0 0.0 0.0 0.6 0.8 0.0 1.6 0.0 0.2 0.0
## [15] 0.0 3.0 0.4 0.0 0.0 5.2
....

te_target <- ds[test,][[target]] %T>%
             {head(., 20) %>% print()}

##  [1] no  no  no  no  no  no  yes no  no  no  yes yes no  no
## [15] no  no  no  no  no  yes
....

te_risk   <- ds[test,][[risk]] %T>%
             {head(., 20) %>% print()}

##  [1] 0.2 0.0 0.0 0.0 0.0 0.0 3.0 0.0 0.0 0.0 1.2 5.8 0.0 0.0
## [15] 0.0 0.0 0.0 0.0 0.0 6.2
....
```

3.10 Save the Dataset

Having transformed our dataset in a variety of ways, cleaned it, wrangled it, and added new variables, we will now save the data and its metadata into a binary RData file. Saving it as a binary compressed file saves both storage space and time on reloading the dataset. Loading a binary dataset is generally faster than loading a CSV file.

```
# Save data into a appropriate folder.

fpath <- "data"

# Timestamp for the dataset.
```

```
dsdate <- "_" %s+% format(Sys.Date(), "%Y%m%d") %T>% print()

## [1] "_20170615"

# Use a fixed timestamp to name our file for convenience here.

dsdate <- "_20170702"

# Filename for the saved dataset.

dsfile <- dsname %s+% dsdate %s+% ".RData"

# Full path to the dataset.

fpath %>%
  file.path(dsfile) %T>%
  print() ->
dsrdata

## [1] "data/weatherAUS_20170702.RData"

# Save relevant R objects to the binary RData file.

save(ds, dsname, dspath, dsdate, nobs,
     vars, target, risk, id, ignore, omit,
     inputi, inputs, numi, numc, cati, catc,
     form, seed, train, validate, test,
     tr_target, tr_risk, va_target, va_risk, te_target, te_risk,
     file=dsrdata)

# Check the resulting file size in bytes.

file.size(dsrdata) %>% comma()

## [1] "5,568,471"
```

Notice that in addition to the dataset (`ds`) we also store the collection of ***metadata***. This begins with items such as the name of the dataset, the source file path, the date we obtained the dataset, the number of observations, the variables of interest, the target variable, the name of the risk variable (if any), the identifiers, the variables to ignore and observations to omit. We continue with the indicies of the input variables and their names, the indicies of the numeric variables and their names, and the indicies of the categoric variables and their names.

Each time we wish to use the dataset we can now simply base::load() it into R. The value that is invisibly returned by base::load() is a vector naming the R objects loaded from the binary RData file.

```
load(dsrdata) %>% print()

##  [1] "ds"         "dsname"     "dspath"     "dsdate"
##  [5] "nobs"       "vars"       "target"     "risk"
##  [9] "id"         "ignore"     "omit"       "inputi"
## [13] "inputs"     "numi"       "numc"       "cati"
## [17] "catc"       "form"       "seed"       "train"
## [21] "validate"   "test"       "tr_target"  "tr_risk"
## [25] "va_target"  "va_risk"    "te_target"  "te_risk"
```

A call to base::load() returns its result invisibly since we are primarily interested in its side-effect. The side-effect is to read the R binary data from storage and to make it available within our current R session.

3.11 A Template for Data Preparation

Throughout this chapter we have worked towards constructing a standard template for ourselves for a data preparation report. Indeed, whenever we begin a new project the template will provide a useful starting point. The introduction of generic variables facilitates this approach to quickly begin any new analysis.

A template based on this chapter for data preparation is available from https://essentials.togaware.com. There we can find a template that will continue to be refined over time and will incorporate improvements and advanced techniques that go beyond what has been presented here. An automatically derived version including just the R code is available there together with the LaTeX compiled PDF version.

Notice that we would not necessarily perform all of the steps we have presented in this chapter. Normalizing the variable names or imputing imputing missing values, omitting observations with missing values, and so on may well depend on the context of the analysis. We will pick and choose as is appropriate to our situation and specific datasets. Also, some data-specific transformations are not included in the template and there may be other transforms we need to perform that we have not covered here.

3.12 Exercises

Exercise 3.1 Exploring the Weather

We have worked with the Australian **weatherAUS** dataset throughout this chapter. For this exercise we will explore the dataset further.

1. Create a data preparation script beginning with the template available from `https://essentials.togaware.com` and replicate the data processing performed in this chapter.

2. Investigate the `dplyr::group_by()` and `dplyr::summarise()` functions, combined through a pipeline using `magrittr::%>%` to identify regions with considerable variance in their weather observations. The use of `stats::var()` might be a good starting point.

Exercise 3.2 Understanding Ferries

A dataset of ferry crossings on Sydney Harbour is available as `https://essentials.togaware.com/ferry.csv`.[*] We will use this dataset to exercise our data template.

[*] The original source of the Ferry dataset is `http://www.bts.nsw.gov.au/Statistics/Ferry/default.aspx?FolderID=224`. The dataset is available under a Creative Commons Attribution (CC BY 3.0 AU) license.

1. Create a data preparation script beginning with the template available from `https://essentials.togaware.com/data.R`.

2. Change the sample source dataset within the template to download the ferry dataset into R.

3. Rename the variables to become normalized variable names.

4. Create two new variables from `sub_route`, called `origin` and `destination`.

5. Convert dates as appropriate and convert other character variables into factors where it is sensible to do so.

6. Work through the template to prepare and then explore the dataset to identify any issues and to develop initial observations.

Create a report on your investigations and share a narrative to communicate your discoveries from the dataset.

4

Visualising Data

One of the most important tasks for any data scientist is to visualise data. Presenting data visually will often lead to new insights and discoveries, as well as providing clear evidence of any issues with the data. A visual presentation is also often the most effective means for communicating insight to the business problem owners.

R offers a comprehensive suite of tools to visualise data, with ggplot2 (Wickham and Chang, 2016) being dominate amongst them. The ggplot2 package implements a grammar for writing sentences describing the graphics. Using this package we construct a plot beginning with the dataset and the aesthetics (e.g., x-axis and y-axis) and then add geometric elements, statistical operations, scales, facets, coordinates, and numerous other components.

In this chapter we explore data using ggplot2 to gain new insights into our data. The package provides an extensive collection of capabilities offering an infinite variety of visual possibilities. We will present some basics as a launching pad for plotting data but note that further opportunities abound and are well covered in many other resources and online.

Packages used in this chapter include GGally (Schloerke *et al.*, 2016), RColorBrewer (Neuwirth, 2014), dplyr (Wickham *et al.*, 2017a), ggplot2 (Wickham and Chang, 2016), gridExtra (Auguie, 2016), lubridate (Grolemund *et al.*, 2016), magrittr (Bache and Wickham, 2014), randomForest (Breiman *et al.*, 2015), rattle (Williams, 2017), scales (Wickham, 2016), stringi (Gagolewski *et al.*, 2017), and stringr (Wickham, 2017a).

```
# Load required packages from local library into the R session.

library(GGally)        # Parallel coordinates.
library(RColorBrewer) # Choose different colors.
library(dplyr)         # Data wrangling.
```

```
library(ggplot2)      # Visualise data.
library(gridExtra)    # Layout multiple plots.
library(lubridate)    # Dates and time.
library(magrittr)     # Pipelines for data processing.
library(randomForest) # Deal with missing data.
library(rattle)       # weatherAUS dataset and normVarNames().
library(scales)       # Include commas in numbers.
library(stringi)      # String concat operator %s+%.
library(stringr)      # Strings: str_replace().
```

4.1 Preparing the Dataset

We use the modestly large **weatherAUS** dataset from rattle (Williams, 2017) to illustrate the capabilities of ggplot2. For plots that generate large images we might use random subsets of the same dataset to allow replication in a timely manner. We begin by loading the dataset saved through the template we developed in Chapter 3.

```
# Build the filename used to previously store the data.

fpath  <- "data"
dsname <- "weatherAUS"
dsdate <- "_20170702"
dsfile <- dsname %s+% dsdate %s+% ".RData"

fpath %>%
  file.path(dsfile) %>%
  print() ->
dsrdata

## [1] "data/weatherAUS_20170702.RData"

# Load the R objects from file and list them.

load(dsrdata) %>% print()

##  [1] "ds"        "dsname"    "dspath"    "dsdate"
##  [5] "nobs"      "vars"      "target"    "risk"
```

```
##  [9] "id"         "ignore"     "omit"       "inputi"
## [13] "inputs"     "numi"       "numc"       "cati"
## [17] "catc"       "form"       "seed"       "train"
## [21] "validate"   "test"       "tr_target"  "tr_risk"
## [25] "va_target"  "va_risk"    "te_target"  "te_risk"
```

We will perform missing value imputation but note that this is not something we should be doing lightly (inventing new data). We do so here simply to avoid warnings that would otherwise advise us of missing data when using ggplot2. Note the use of `randomForest::na.roughfix()` to perform missing value imputation as discussed in Chapter 3. Alternatively we could remove observations with missing values using `stats::na.omit()`.

```
# Count the number of missing values.

ds[vars] %>% is.na() %>% sum() %>% comcat()

## 278,293

# Impute missing values.

ds[vars] %<>% na.roughfix()

# Confirm that no missing values remain.

ds[vars] %>% is.na() %>% sum() %>% comcat()

## 0
```

We are now in a position to visually explore our dataset. We begin with a textual `dplyr::glimpse()` of the dataset for reference.

```
glimpse(ds)

## Observations: 134,776
## Variables: 27
## $ date             <date> 2008-12-01, 2008-12-02, 2008-12-0...
## $ location         <fctr> albury, albury, albury, albury, a...
## $ min_temp         <dbl> 13.4, 7.4, 12.9, 9.2, 17.5, 14.6, ...
## $ max_temp         <dbl> 22.9, 25.1, 25.7, 28.0, 32.3, 29.7...
```

```
## $ rainfall         <dbl> 0.6, 0.0, 0.0, 0.0, 1.0, 0.2, 0.0,...
## $ evaporation      <dbl> 4.8, 4.8, 4.8, 4.8, 4.8, 4.8, 4.8,...
## $ sunshine         <dbl> 8.5, 8.5, 8.5, 8.5, 8.5, 8.5, 8.5,...
## $ wind_gust_dir    <ord> w, wnw, wsw, ne, w, wnw, w, w, nnw...
## $ wind_gust_speed <dbl> 44, 44, 46, 24, 41, 56, 50, 35, 80...
## $ wind_dir_9am     <ord> w, nnw, w, se, ene, w, sw, sse, se...
## $ wind_dir_3pm     <ord> wnw, wsw, wsw, e, nw, w, w, w, nw,...
## $ wind_speed_9am   <int> 20, 4, 19, 11, 7, 19, 20, 6, 7, 15...
## $ wind_speed_3pm   <dbl> 24, 22, 26, 9, 20, 24, 24, 17, 28,...
## $ humidity_9am     <int> 71, 44, 38, 45, 82, 55, 49, 48, 42...
## $ humidity_3pm     <int> 22, 25, 30, 16, 33, 23, 19, 19, 9,...
## $ pressure_9am     <dbl> 1007.7, 1010.6, 1007.6, 1017.6, 10...
## $ pressure_3pm     <dbl> 1007.1, 1007.8, 1008.7, 1012.8, 10...
## $ cloud_9am        <dbl> 8, 5, 5, 5, 7, 5, 1, 5, 5, 5, 5, 8...
## $ cloud_3pm        <dbl> 5, 5, 2, 5, 8, 5, 5, 5, 5, 5, 5, 8...
## $ temp_9am         <dbl> 16.9, 17.2, 21.0, 18.1, 17.8, 20.6...
## $ temp_3pm         <dbl> 21.8, 24.3, 23.2, 26.5, 29.7, 28.9...
## $ rain_today       <fctr> no, no, no, no, no, no, no, no, n...
## $ risk_mm          <dbl> 0.0, 0.0, 0.0, 1.0, 0.2, 0.0, 0.0,...
## $ rain_tomorrow    <fctr> no, no, no, no, no, no, no, no, y...
## $ year             <fctr> 2008, 2008, 2008, 2008, 2008, 200...
## $ season           <fctr> summer, summer, summer, summer, s...
## $ cluster          <fctr> area2, area2, area2, area2, area2...
```

4.2 Scatter Plot

Our first plot is a simple scatter plot which displays points scattered over a plot. A difficulty with scatter plots (and indeed with many types of plots) is that for large datasets we end up with rather dense plots. For illustrative purposes we will identify a random subset of just 1,000 observations to plot. We thus avoid filling the plot completely with points as might otherwise happen as in Figure 2.3.

```
ds %>%
  nrow() %>%
  sample(1000) ->
sobs
```

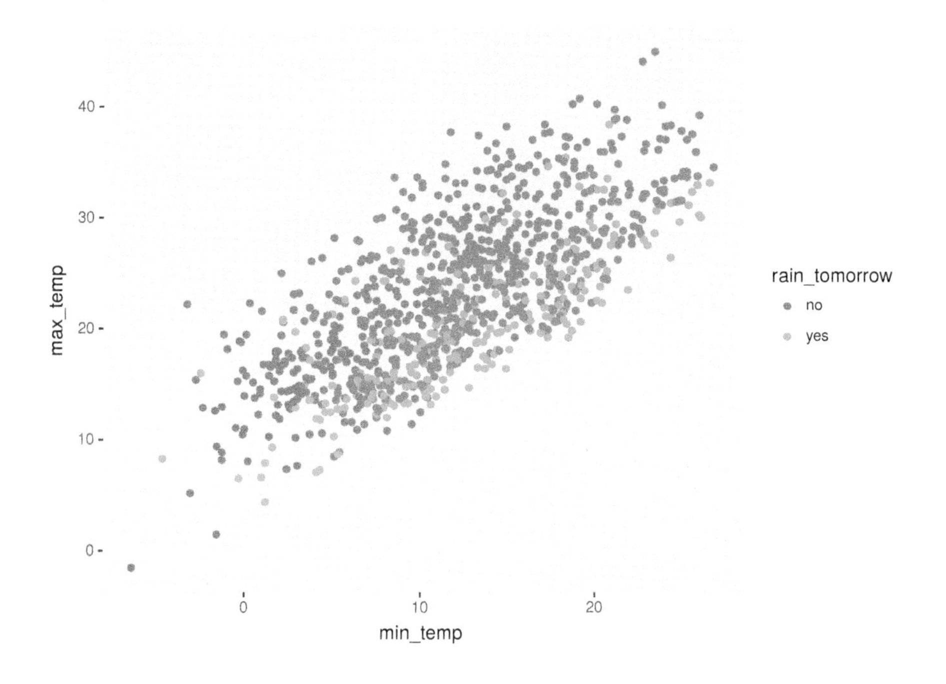

Figure 4.1: *Scatter plot of the weatherAUS dataset.*

We are now ready to generate the plot. We choose just the random sample of rows whose indices are stored in the variable `sobs`. This subset is piped through to `ggplot2::ggplot()` which initialises the plot. To the plot we add points using `ggplot2::geom_point()`. The resulting plot is displayed in Figure 4.1.

```
ds %>%
  extract(sobs,) %>%
  ggplot(aes(x=min_temp, y=max_temp, colour=rain_tomorrow)) +
  geom_point()
```

The call to `ggplot2::ggplot()` includes as its argument the aesthetics of the plot. We identify for the `x=` axis the variable `min_temp` and for the `y=` axis the variable `max_temp` as the y-axis. In addition to minimally identifying the x and y mapping we add a `colour=` option to distinguish between those days where `rain_tomorrow` is true from those where it is false.

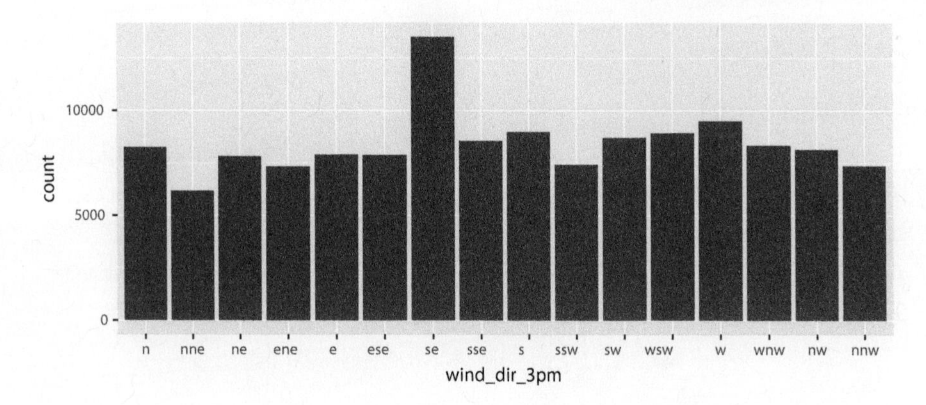

Figure 4.2: *Bar chart showing the relative occurrence of different wind directions as recorded at 3pm.*

Having set up the aesthetics of the plot we can add a graphical layer. The simplest is just to plot the points (x, y) coloured appropriately. All that is required is to add (i.e., "+") a call to `ggplot2::geom_point()`.

4.3 Bar Chart

Another common plot is the bar chart which displays the count of observations using bars. Such plots are generated by **ggplot2** using `ggplot2::geom_bar()`. Figure 4.2 is generated with:

```
ds %>%
  ggplot(aes(x=wind_dir_3pm)) +
  geom_bar()
```

Here we only require an x-axis to be specified which in our example is `wind_dir_3pm`. To the base plot created by `ggplot2::ggplot()` we add a so-called bar geometric to construct the required plot. The resulting plot shows the frequency of the levels of the categoric variable `wind_dir_3pm` across the whole dataset.

4.4 Saving Plots to File

Generating a plot is one thing but we will want to make use of the plot possibly in multiple ways. Once we have a plot displayed we can save the plot to file quite simply using `ggplot2::ggsave()`. The format is determined automatically by the name of the file to which we save the plot. Here, for example, we save the plot as a PDF that we might include in other documents or share with a colleague for discussion.

```
ggsave("barchart.pdf", width=11, height=7)
```

Notice the use of `width=` and `height=`. The default values are those of the current plotting window so for saving the plot we have specified a particular width and height. By trial and error or by experience we have found the proportions used here to suit our requirements.

There is some art required in choosing a good width and height as we discuss in Chapter 10. By increasing the height or width any text that is displayed on the plot essentially stays the same size. Thus by increasing the plot size the text will appear smaller. By decreasing the plot size the text becomes larger. Some experimentation is often required to get the right size for any particular purpose.

4.5 Adding Spice to the Bar Chart

A bar chart can be enhanced in many ways to demonstrate different characteristics of the data. A stacked bar chart is commonly used to identify the distribution of the observations over another variable, like the target variable. Our target is **rain_tomorrow** and we obtain a stacked bar chart by filling the bars with colour based on this variable using `fill=`. This is implemented as follows with the result displayed in Figure 4.3.

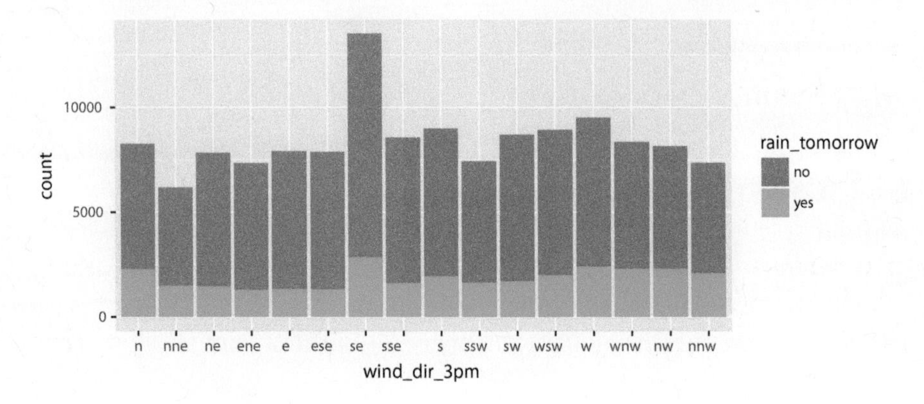

Figure 4.3: *Stacked bar chart.*

```
ds %>%
  ggplot(aes(x=wind_dir_3pm, fill=rain_tomorrow)) +
  geom_bar()
```

There are many options available to tune how we present our plots using ggplot2. In a similar way to building pipelines of functions to achieve our data processing as we saw in Chapters 2 and 3, we build our plots incrementally. We will illustrate a number of options in the following codes as we build a more interesting presentation of the data as in Figure 4.4. Be sure to replicate the plot by adding one line of the following code at a time and studying its impact before moving on to the next line/option/layer. We detail some of the options below.

```
blues2 <- brewer.pal(4, "Paired")[1:2] %T>% print()

## [1] "#A6CEE3" "#1F78B4"

ds$location %>%
  unique() %>%
  length() %T>%
  print() ->
num_locations

## [1] 49
```

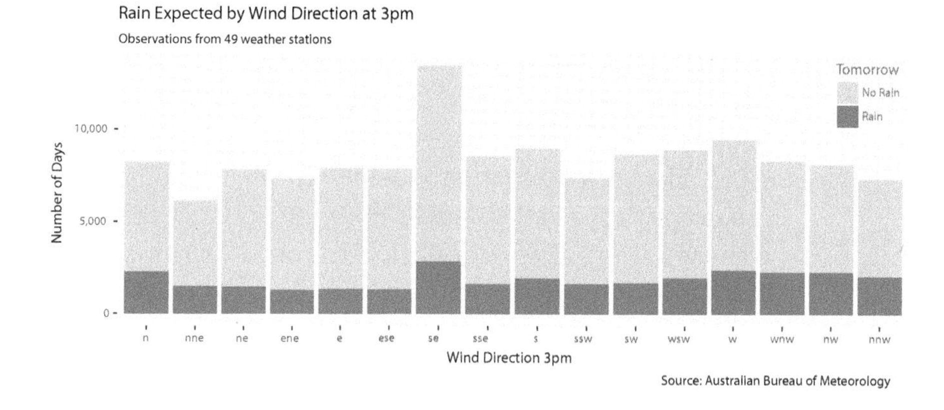

Figure 4.4: *A decorated stacked bar chart.*

```
ds %>%
  ggplot(aes(x=wind_dir_3pm, fill=rain_tomorrow)) +
  geom_bar() +
  scale_fill_manual(values = blues2,
                    labels = c("No Rain", "Rain")) +
  scale_y_continuous(labels=comma) +
  theme(legend.position   = c(.95, .85),
        legend.title      = element_text(colour="grey40"),
        legend.text       = element_text(colour="grey40"),
        legend.background = element_rect(fill="transparent")) +
  labs(title    = "Rain Expected by Wind Direction at 3pm",
       subtitle = "Observations from " %s+%
                  num_locations %s+%
                  " weather stations",
       caption  = "Source: Australian Bureau of Meteorology",
       x        = "Wind Direction 3pm",
       y        = "Number of Days",
       fill     = "Tomorrow")
```

The most obvious change is to the colouring which is supported by `RColorBrewer::brewer.pal()`. This is used to generate a dark/light pair of colours that are softer in presentation. We use only the first two colours from the generated palette of four colours.

In reviewing the original plot we might also notice that the y scale is in the thousands and yet no comma is used to emphasise that. It is always a good idea to include commas in large numbers to denote the thousands and avoid misreading. We do so by specifying that the y scale labels should use scales::**comma()**. We also include more informative labels through the use of ggplot2::**labs()**.

The new plot is arguably more appealing and marginally more informative than the original. We might note though that the most interesting question we can ask in relation to the data behind this plot is whether there is any differences in the distribution between rain and no rain across the different wind directions. Our chosen plot does not facilitate answering this question. A simple change to ggplot2::**geom_bar()** by adding position="fill" results in Figure 4.5. Notice that we moved the legend back to its original position as there is now no empty space within the plot.

```
ds %>%
  ggplot(aes(x=wind_dir_3pm, fill=rain_tomorrow)) +
  geom_bar(position="fill") +
  scale_fill_manual(values = blues2,
                    labels = c("No Rain", "Rain")) +
  scale_y_continuous(labels=comma) +
  theme(legend.title      = element_text(colour="grey40"),
        legend.text       = element_text(colour="grey40"),
        legend.background = element_rect(fill="transparent")) +
  labs(title    = "Rain Expected by Wind Direction at 3pm",
       subtitle = "Observations from " %s+%
                  num_locations %s+%
                  " weather stations",
       caption  = "Source: Australian Bureau of Meteorology",
       x        = "Wind Direction 3pm",
       y        = "Number of Days",
       fill     = "Tomorrow")
```

Observe now that indeed wind direction appears to have an influence on whether it rains the following day. Northerly winds appear to have a higher proportion of following days on which it rains. Any such observation requires further statisical confirmation to be sure.

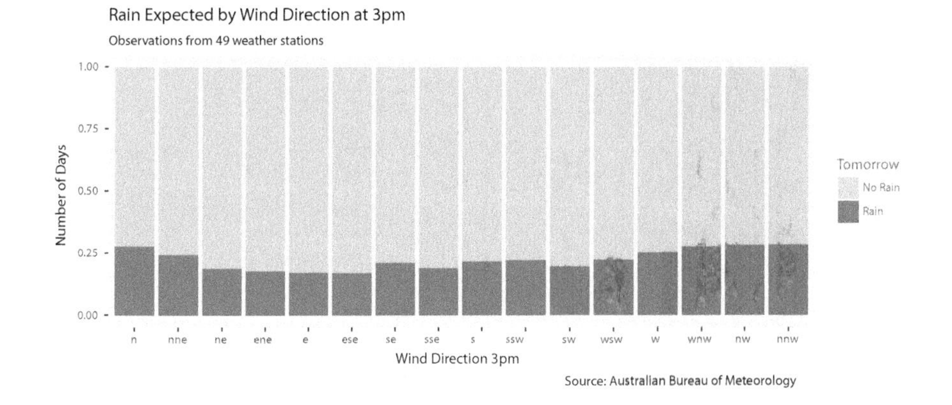

Figure 4.5: *A decorated stacked filled bar chart.*

4.6 Alternative Bar Charts

There are a variety of options available to tune how we present a bar chart using ggplot2::**ggplot()**. We illustrate here dealing with a few more bars in the plot by visualising the mean temperature at 3pm for locations in the dataset.

We start with the plot shown in Figure 4.6. The plot has x= **location** and y= **temp_3pm** with a fill= set to the **location**. The choice of fill simply results in each bar being rendered with a different colour but adds no real value to the plot other than, arguably, its appeal. To this basic setup we add a stat= **summary** and calculate the bars as the mean value of the **y** axis variable using fun.y=.

```
ds %>%
  ggplot(aes(x=location, y=temp_3pm, fill=location)) +
  geom_bar(stat="summary", fun.y="mean") +
  theme(legend.position="none")
```

There are 49 locations represented in the dataset resulting in quite a clutter of location names along the x axis. The obvious solution is to rotate the labels which we achieve by modifying the ggplot2::**theme()** through setting the axis.text= to be rotated by an angle= of 90°. The result is shown in Figure 4.7.

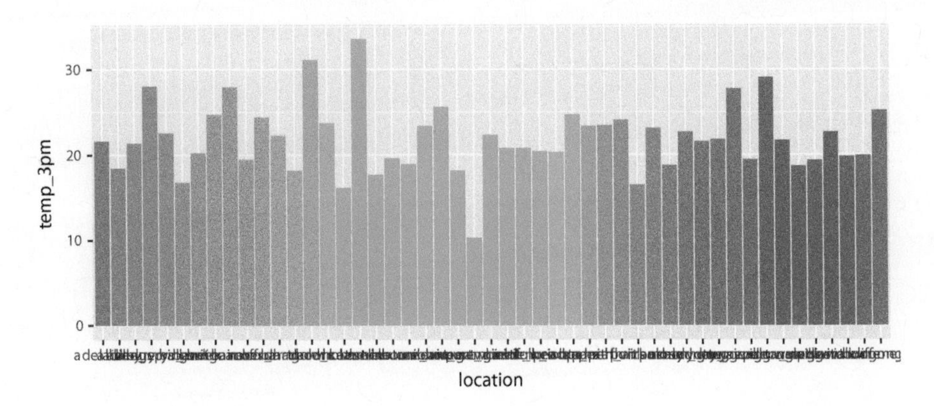

Figure 4.6: *Multiple bars with overlapping labels.*

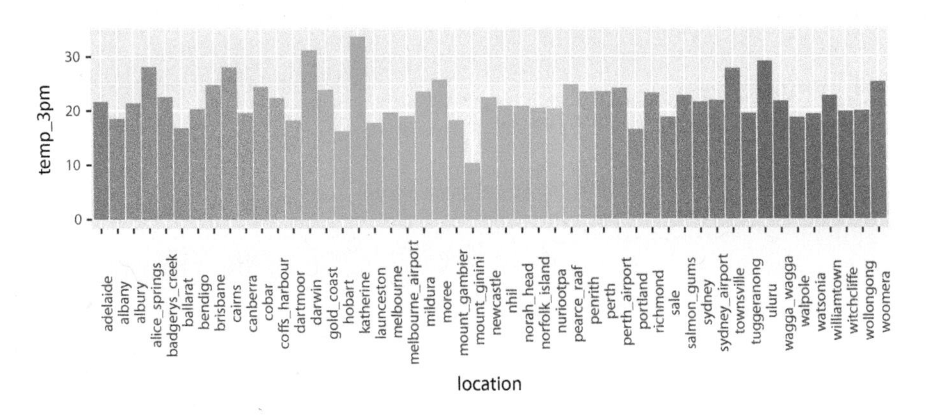

Figure 4.7: *Rotating labels in a plot.*

```
ds %>%
  ggplot(aes(location, temp_3pm, fill=location)) +
  geom_bar(stat="summary", fun.y="mean") +
  theme(legend.position="none") +
  theme(axis.text.x=element_text(angle=90))
```

Instead of flipping the labels we could flip the coordinates and produce a horizontal bar chart as in Figure 4.8. Rotating the plot allows more bars to be readily added down the page than we might across the page. It is also easier for us to read the labels left to right rather than bottom up. However the plot is less compact.

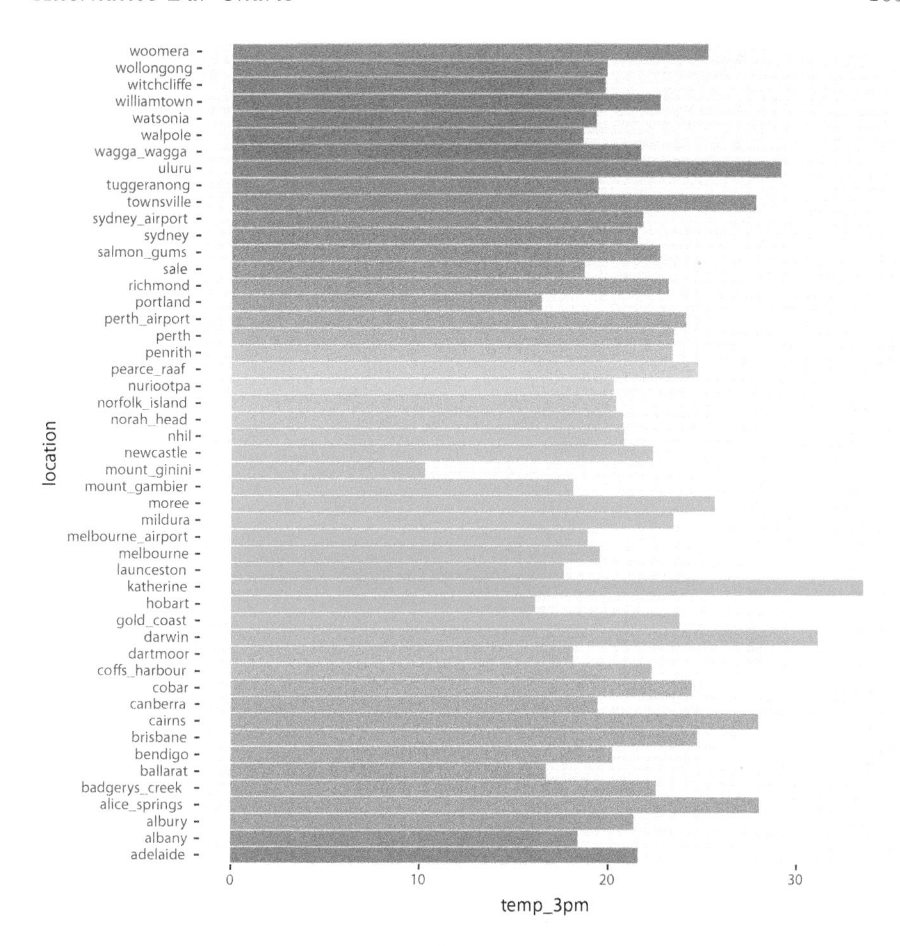

Figure 4.8: *Rotating the plot.*

```
ds %>%
  ggplot(aes(location, temp_3pm, fill=location)) +
  geom_bar(stat="summary", fun.y="mean") +
  theme(legend.position="none") +
  coord_flip()
```

We would also naturally be inclined to expect the labels to appear in alphabetic order rather than the reverse as it appears by default. One approach is to reverse the order of the levels in the original dataset. We can use `dplyr::mutate()` within a

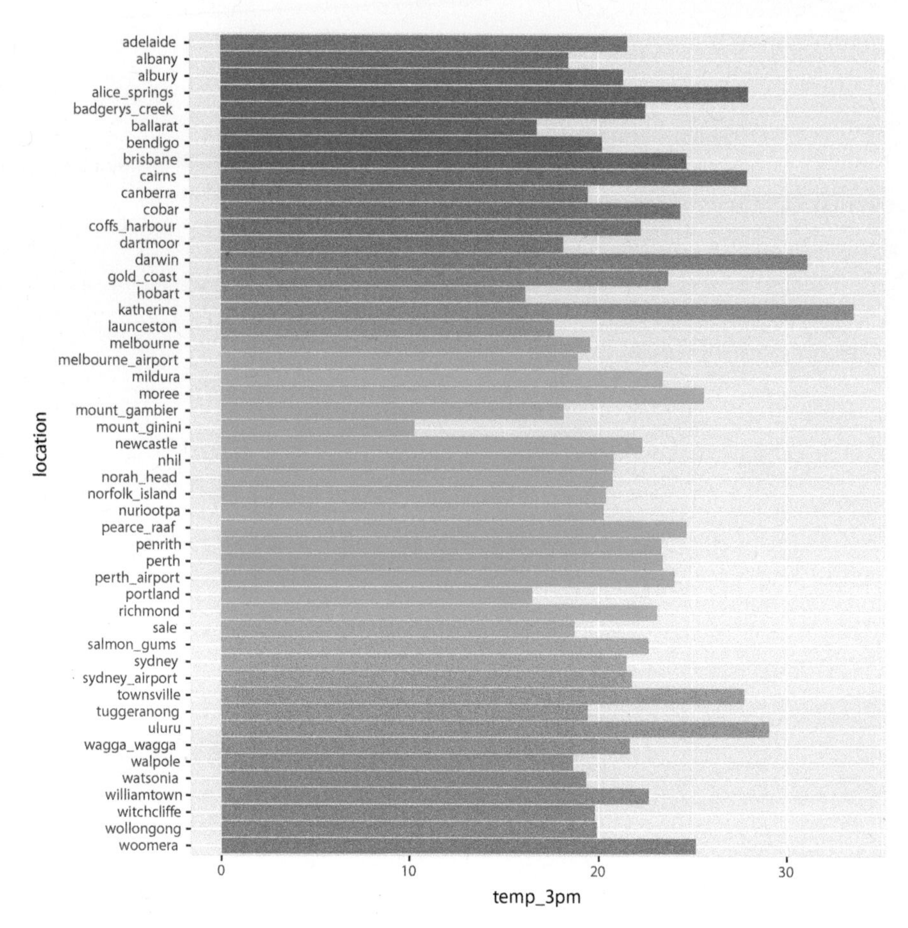

Figure 4.9: *Reordering labels.*

pipeline to temporarily do this and pass the modified dataset on to `ggplot2::ggplot()`. The result can be seen in Figure 4.9.

```
ds$location %>%
  levels() %>%
  rev() ->
loc

ds %>%
  mutate(location=factor(location, levels=loc)) %>%
  ggplot(aes(location, temp_3pm, fill=location)) +
  stat_summary(fun.y="mean", geom="bar") +
  theme(legend.position="none") +
```

```
coord_flip()
```

4.7 Box Plots

A box plot, also known as a box and whiskers plot, is another tool to visualise the distribution of our data. The box plot of Figure 4.10 shows the median value (the midpoint) of the variable as the horizontal line within the box. The box itself contains half the observations with one quarter of the remaining observations shown below and one quarter of the remaining observations shown above the box. Specific points are displayed as outliers at the extremes. An outlier is a value that is some distance from the bulk of the data. We use ggplot2::geom_boxplot() to add a box plot to out canvas.

```
ds %>%
  ggplot(aes(x=year, y=max_temp, fill=year)) +
  geom_boxplot(notch=TRUE) +
  theme(legend.position="none")
```

Here the aesthetics for ggplot2::ggplot() are set with x= year and y= max_temp. Colour is added using fill= year. The colour (arguably) improves the visual appeal of the plot—it conveys little if any information. Since we have included fill= we must also turn off the otherwise included but redundant legend through the ggplot2::theme() with a legend.position= none.

Now that we have presented the data we begin our observations of the data. It is noted that the first and last verticals look different to the others due to the data likely being truncated. Our task is to confirm this indeed is a fact of the data itself. The 2017 data is truncated to the beginning of the year in this particular dataset. It is clear that the year has begun with very hot temperatures, remembering that being from Australia this data is recorded for a summer month.

A variation of the box plot is the violin plot. The violin plot adds information about the distribution of the data. The resulting

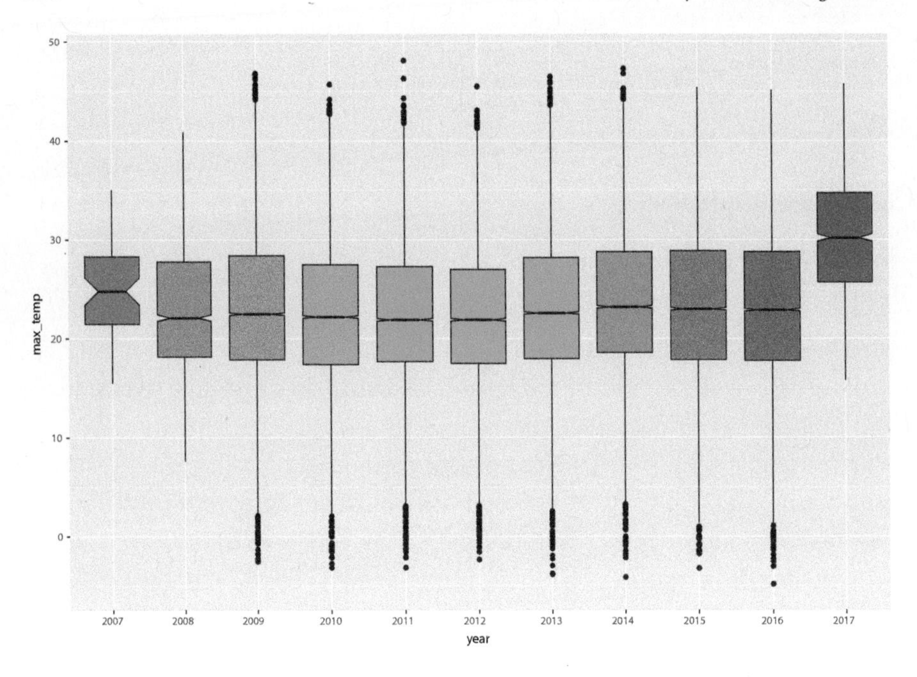

Figure 4.10: *A traditional box and wiskers plot.*

shape often reflects that of a violin. In Figure 4.11 we again use colour to improve the visual appeal of the plot.

```
ds %>%
  ggplot(aes(x=year, y=max_temp, fill=year)) +
  geom_violin() +
  theme(legend.position="none")
```

The amount of information we capture in the plot is increased by overlaying a box plot onto the violin plot. This could result in information overload or else it might convey concisely the story that the data tells. Often we need to make a trade off.

```
ds %>%
  ggplot(aes(x=year, y=max_temp, fill=year)) +
  geom_violin() +
  geom_boxplot(width=.5, position=position_dodge(width=0)) +
  theme(legend.position="none")
```

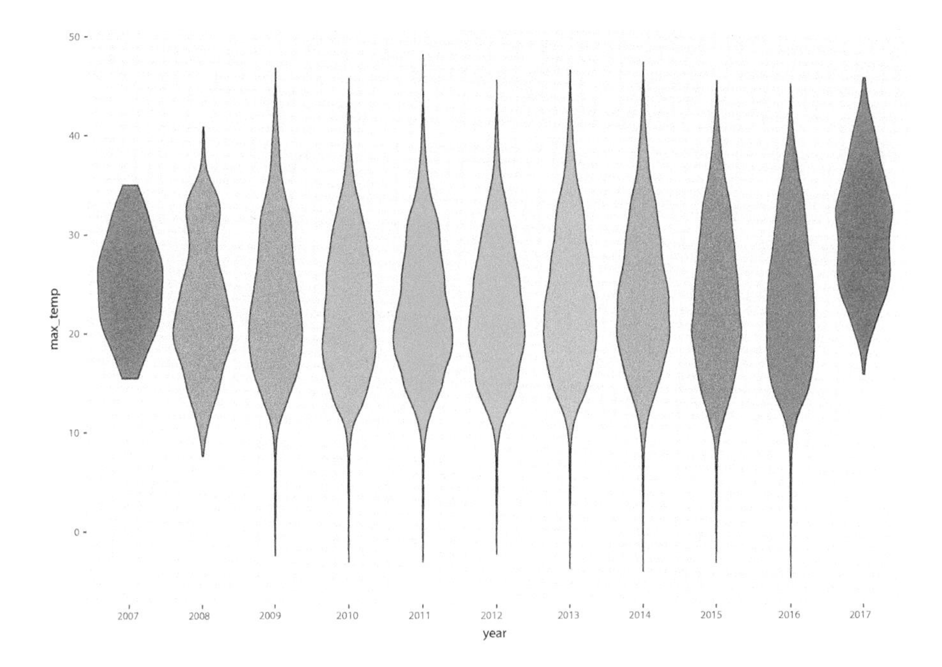

Figure 4.11: *A violin plot.*

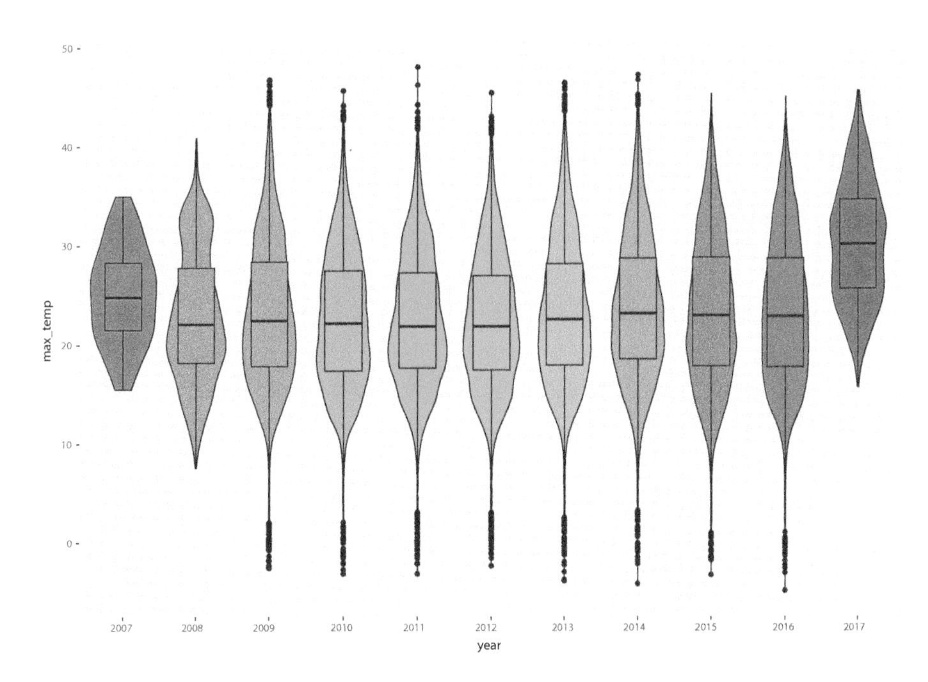

Figure 4.12: *A violin plot with a box plot overlay.*

We can also readily split our plot across locations. The resulting plot (Figure 4.13) is a little crowded but we get an overall view across all of the different weather stations. Notice that we also rotated the x-axis labels so that they don't overlap.

```
ds %>%
  ggplot(aes(x=year, y=max_temp, fill=year)) +
  geom_violin() +
  geom_boxplot(width=.5, position=position_dodge(width=0)) +
  theme(legend.position="none") +
  theme(axis.text.x=element_text(angle=45, hjust=1)) +
  facet_wrap(~location, ncol=5)
```

By creating this plot we can identify further issues exhibited by the data. For example, it is clear that three weather stations have fewer observations than the others as their plots are obviously truncated on the left. This may be a data collection issue or simply that those weather stations are newly installed. The actual reasoning may or may not be important but it is important to question and to understand what we observe from the data. Often we will identify systemic issues with the data that need to be addressed in modelling.

We notice also the issue we identified earlier of apparently few observations before 2009 with only the location `Canberra` having any observations in 2007. In seeking an explanation we come to understand this was simply the nature of how the data was collected from the Bureau. In any modelling over all locations we may decide to eliminate pre-2009 observations.

Further observing the characteristics of the data we note that some locations have quite minimal variation in their maximum temperatures over the years (e.g., Darwin) whilst others can swing between extremes. We also observe that most outliers appear to be at the warmer end of the scale rather than the colder end. There is also a collection of locations which appear to have no outliers. And so on. We are beginning on a journey of discovery—to discover our data—to live and breathe the data (Williams, 2011).

We follow up our observations with modelling to cluster the locations, for example, according to the visual patterns we have just

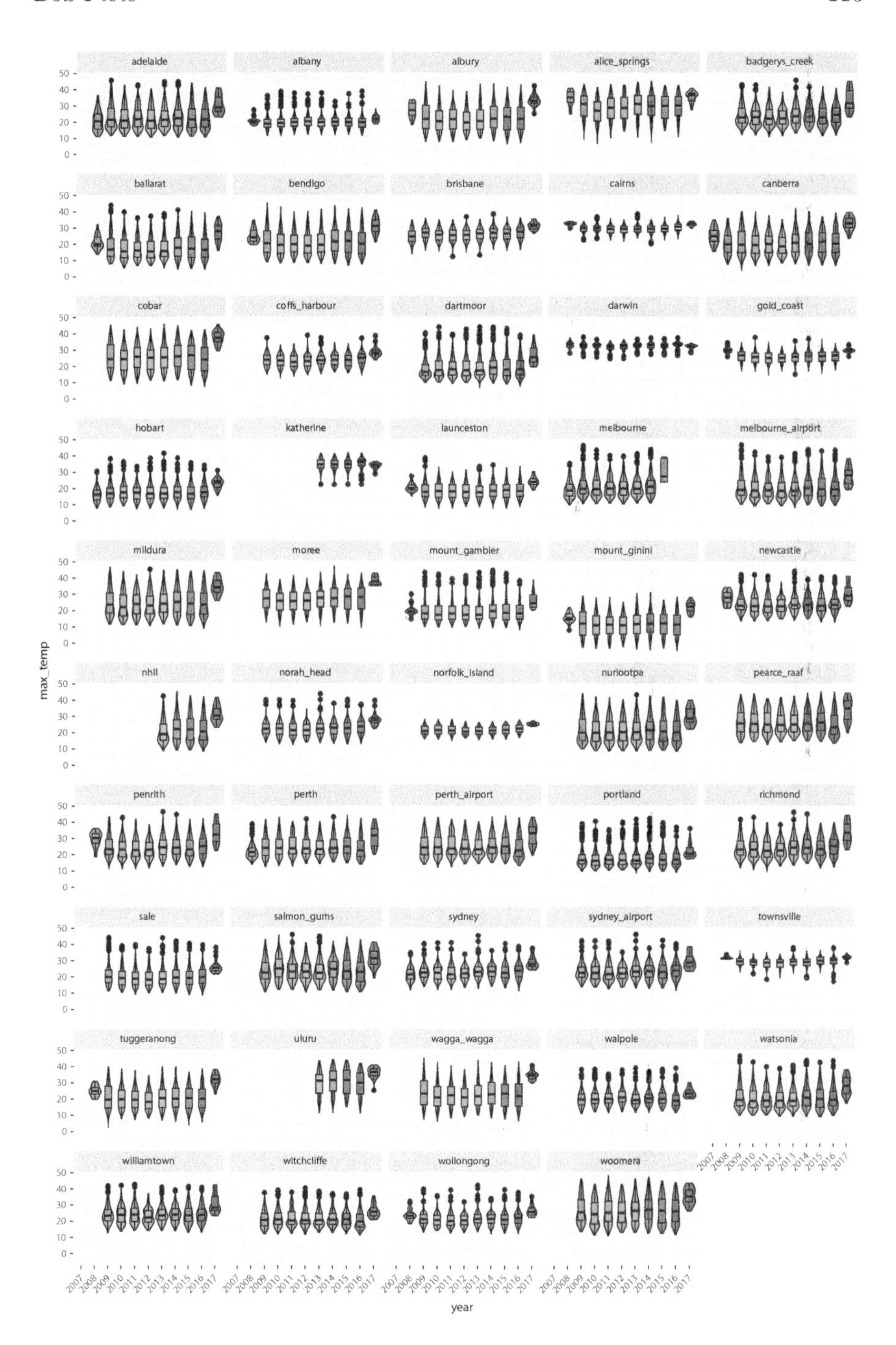

Figure 4.13: *Violin/box plot by location.*

observed. In particular, as our datasets increase in size (this current dataset has just 134,776 observations) we often need to subset the data in various ways in order to visualise the observations.

A cluster analysis provides a simple mechanism to identify groups and allows us to then visualise within those groups. We illustrated the process of a cluster analysis over the numeric variables in Chapter 3. The aim of the cluster analysis, introducing a new feature, was to group the locations so that within a group the locations have similar values for the numeric variables and between the groups the numeric variables are more dissimilar. The resulting feature `cluster` allocated each `location` to a region and we can use that here.

We will construct a plot using exactly the same sequence of commands we used for Figure 4.13. On the principle of avoiding repeating ourselves we create a function to capture the sequence of commands and thus allow multiple plots to be generated by simply calling the function rather than writing out the code each time.

```
# For convenience define a function to generate the plot.

myplot <- function(ds, n)
{
  ds %>%
    filter(cluster==n) %>%
    ggplot(aes(x=year, y=max_temp, fill=year)) +
    geom_violin() +
    geom_boxplot(width=.5, position=position_dodge(width=0)) +
    theme(legend.position="none") +
    theme(axis.text.x=element_text(angle=45, hjust=1)) +
    facet_wrap(~location)
}
```

We can now plot specific clusters with the results in Figures 4.14 and 4.15.

```
# Visualise specific cluster of locations.

myplot(ds, "area4")
```

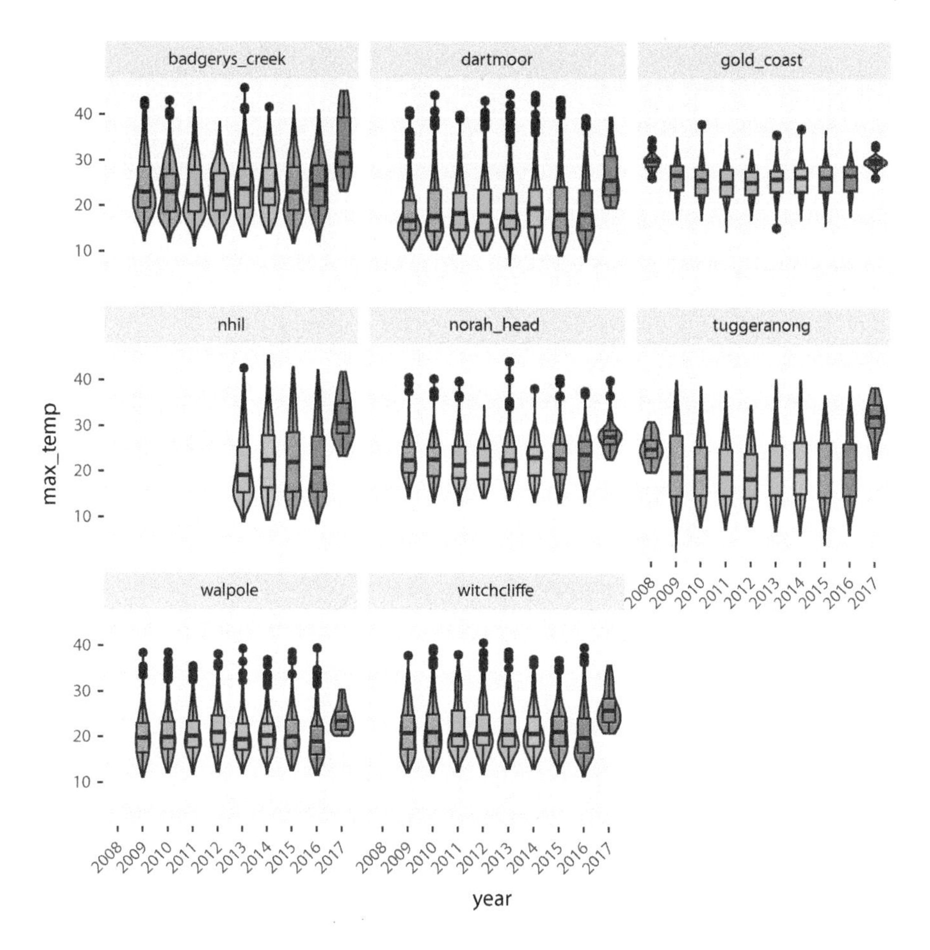

Figure 4.14: *Visualise the first set of clustered locations.*

```
# Visualise specific cluster of locations.

myplot(ds, "area5")
```

Being a grammar for graphics, or essentially a language in which to express our graphics, there is an infinite variety of possibilities with ggplot2. We will see further examples in the following chapters and extensively on the Internet.

As we explore the dataset and observe and question the characteristics of the dataset, we will move from visual observation to programmatic exploration and statistical confirmation.

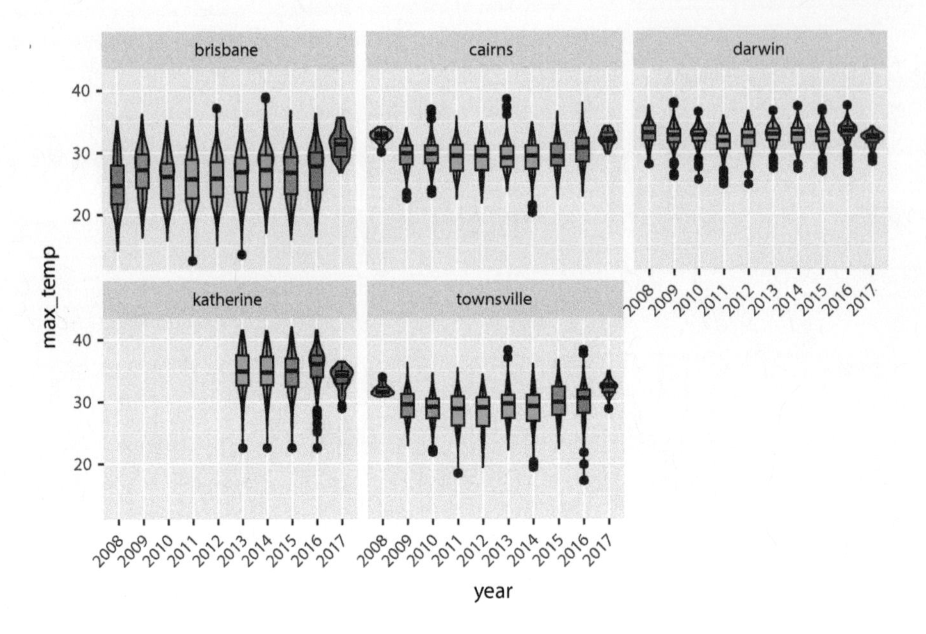

Figure 4.15: *Visualise the second set of clustered locations.*

4.8 Exercises

Exercise 4.1 Dashboard

Your computer collects information about its usage as log files. Identify some log files that you have access to on your own computer. Build a dashboard using ggplot2 to provide informative plots of the information contained in the log data. Include at least a bar plot and line chart, illustrating activity over time. Explore options for how you might turn these plots into a live dashboard.

Exercise 4.2 Visualising Ferries

In the Chapter 3 exercises we introduced the ferry dataset. Continue your observations of that dataset through a visual exploration. Build on the narrative developed there to produce an enhanced narrative that includes visualisations that support the narrative.

5

Case Study: Australian Ports

Our understanding of programming and the particularities of a language like R is enhanced through programming by example. This is an effective approach to learning the tools of data science. Building on the foundations so far, which have introduced the basics of loading, wrangling, exploring and visualising data, we now present as an example the data science analysis behind an information sheet published by the Australian government. Our focus is on basic data wrangling and data visualisation to gain insights which lead to a narrative. Visual presentations are extremely effective and we demonstrate a variety of useful presentations used in the information sheet using R and ggplot2 (Wickham and Chang, 2016).

The data used is based on a 2014 study by the Australian Bureau of Infrastructure, Transport and Regional Economics titled *Ports: Jobs Generation in a Context of Regional Development*[*] (BITRE, 2014). The data is not large, emphasising that data science is about the analysis of data to provide insights to support our narratives and not always about big data. The data was originally sourced from the Australian Bureau of Statistics and is approximated here from the published report for illustrative purposes. The government study analysed the potential of sea ports across Australia for future development and considered the impact on jobs for the surrounding regions. Many plots used in the information sheet were generated using ggplot2—we illustrate here code to reproduce these plots.

The information sheet presents a narrative supported by the data and the plots. The narrative is comprehensive and we do not attempt to retell it here. Rather the aim is to demonstrate the

[*] http://www.bitre.gov.au/publications/2014/files/is_056.pdf.

process of analysing the data and provide the practical tools used to support the narrative. The information sheet is an illustration of the output from the data scientist.

Packages used in this chapter include directlabels (Hocking, 2017), dplyr (Wickham *et al.*, 2017a), ggplot2 (Wickham and Chang, 2016), magrittr (Bache and Wickham, 2014), rattle (Williams, 2017), readr (Wickham *et al.*, 2017b), readxl (Wickham and Bryan, 2017), scales (Wickham, 2016), stringi (Gagolewski *et al.*, 2017), stringr (Wickham, 2017a) and tidyr (Wickham, 2017b),

```
# Load required packages from local library into R session.

library(directlabels) # Dodging labels for ggplot2.
library(dplyr)        # Data wrangling.
library(ggplot2)      # Visualise data.
library(magrittr)     # Use pipelines for data processing.
library(rattle)       # normVarNames().
library(readr)        # Modern data reader.
library(readxl)       # Read Excel spreadsheets.
library(scales)       # Include commas in numbers.
library(stringi)      # The string concat operator %s+%.
library(stringr)      # String manpiulation.
library(tidyr)        # Tidy the dataset.
```

5.1 Data Ingestion

The data for this chapter is collected together into an Excel spreadsheet.[*] We've captured the data in a typical scenario that is common for those augmenting their analysis using Excel with analyses using R. It is increasingly important that we are able to readily access such data and the `readxl::read_excel()` function facilitates the data ingestion.

The dataset can be readily downloaded to the local computer using `utils::download.file()` if we are connected to the In-

[*] `http://essentials.togaware.com/ports.xlsx`.

ternet. We do so here saving the downloaded file into a location recorded in `dspath`.

```
# Idenitfy the source location of the dataset.

dsurl  <- "http://essentials.togaware.com/ports.xlsx"

# Local location of the downloaded file.

dspath <- "cache/ports.xlsx"

# Download the file from the Internet.

download.file(dsurl, destfile=dspath, mode="wb")
```

All of the tables that we are interested in are collected together within the first sheet of the spreadsheet. It is not well organised but purposely reflects how we often receive data—data within spreadsheets are not always well designed. Indeed the spreadsheet here is rather haphazard as a typical user might quickly bring together to create basic graphics in Excel.

We will extract the various tables of data from the spreadsheet with specific cell ranges for each plot. We could use Excel or Libre Office Calc to manipulate the data and improve the structure to facilitate data extraction in R but instead we will do all of the data wrangling in R itself.

A potential advantage of not modifying the spreadsheet structure is that the spreadsheet owner may provide an update of the numbers within the tables and we would not want to manually restructure their sheet each time. Of course we might suggest to them to restructure their source spreadsheet but we assume not for now—such advise is not always welcome regardless of how wise it is. Nonetheless we retain the principle of minimising manual effort and the principle of automating processes.

We will load the spreadsheet into R and explore it to determine the appropriate cell ranges for the tables of data we are interested in. A spreadsheet may contain multiple sheets and we are able to list them using `readxl::excel_sheets()`.

```
# List the populated sheets.

excel_sheets(path=dspath)

## [1] "Sheet1"
```

A single sheet is found and it has its default name of `Sheet1`. This is the sheet we read into R.

```
# Ingest the dataset.

ports <- read_excel(path=dspath, sheet=1, col_names=FALSE)

# Prepare the dataset for usage with our template.

dsname <- "ports"
ds     <- get(dsname) %T>% print()

## # A tibble: 117 x 18
##                     X__1      X__2     X__3   X__4    X__5
##                    <chr>     <chr>    <chr>  <chr>   <chr>
##  1                     1         2        3      4       5
##  2                  <NA>  Adelaide Brisbane Burnie Dampier
##  3               2011-12      15.5       37      4     176
##  4 AvgAnnualGrowth10yr         4.9        6    3.5     6.5
##  5                  <NA>      <NA>     <NA>   <NA>    <NA>
##  6                 Mixed      Bulk     <NA>   <NA>    <NA>
##  7             Melbourne   Dampier     <NA>   <NA>    <NA>
##  8              Brisbane Gladstone     <NA>   <NA>    <NA>
##  9           Port Kembla Hay Point     <NA>   <NA>    <NA>
## 10             Devonport Newcastle     <NA>   <NA>    <NA>
## # ... with 107 more rows, and 13 more variables: X__6 <chr>,
## #   X__7 <chr>, X__8 <chr>, X__9 <chr>, X__10 <chr>,
## #   X__11 <chr>, X__12 <chr>, X__13 <chr>, X__14 <chr>,
## #   X__15 <chr>, X__16 <chr>, X__17 <chr>, X__18 <chr>
```

We also `base::print()` the dataset here using the tee-pipe operator for an initial view of the data that has been ingested.

We can begin to see the structure of the sheet and the issues we face with wrangling the data into shape. Essentially an exploration of the dataset using Microsoft Excel or Libre Office Calc or with RStudio using `utils::View()` and working with the data owner will guide us in identifying the location of the embedded tables. In the following we will process the various datasets in the same order as the figures in the Australian Government information sheet.

5.2 Bar Chart: Value/Weight of Sea Trade

The first two figures of interest in the information sheet (Figures 2 and 3) show the dollar value and the weight of imports and exports through sea ports over a period of 11 years. As the spreadsheet contains multiple tables in the sheet that we have ingested, our task is to extract the table of interest into a data frame containing just the data from that table.

The table is found to range over rows 71 to 93 and consist of the first four columns. We ignore the first row of column names.

```
# Confirm the row and column span for the table of interest.

ds[72:93, 1:4]

## # A tibble: 22 x 4
##        X__1       X__2   X__3   X__4
##       <chr>      <chr>  <chr>  <chr>
##  1 2001-02 Australia  99484  85235
##  2    <NA>   17 Ports  84597  83834
##  3 2002-03 Australia  93429  94947
##  4    <NA>   17 Ports  80170  93566
##  5 2003-04 Australia  89303  93467
##  6    <NA>   17 Ports  76163  92045
##  7 2004-05 Australia 106341 108923
##  8    <NA>   17 Ports  92091 106860
##  9 2005-06 Australia 130856 122211
## 10    <NA>   17 Ports 112278 118779
## # ... with 12 more rows
```

The data relates to the dollar value of imports and exports through 17 different sea ports around Australia. We need to understand the data through the use of plots but will need to wrangle the data to do so. As a starting point we note that the columns have rather generic names like `X__1` and `X__2`. We can define new column names in consultation with the data owner using magrittr::set_names().

```
# Wrangle the dataset: Rename columns informatively.

ds[72:93, 1:4] %>%
  set_names(c("period", "location", "export", "import"))

## # A tibble: 22 x 4
##     period  location export import
##      <chr>     <chr>  <chr>  <chr>
##  1 2001-02 Australia  99484  85235
##  2    <NA>  17 Ports  84597  83834
##  3 2002-03 Australia  93429  94947
##  4    <NA>  17 Ports  80170  93566
##  5 2003-04 Australia  89303  93467
##  6    <NA>  17 Ports  76163  92045
##  7 2004-05 Australia 106341 108923
....
```

We now delve into the meaning and structure of the dataset. The `location` is either all Australian ports or just the 17 that are of interest for the analysis. These are the 17 largest ports in Australia according to the data owner.

The dollar value of the export and import through either all Australian ports or the 17 largest ports is reported in millions of dollars. Checking the dataset we see that `export` and `import` are currently character data types rather than the numeric values we would expect. We can fix that quite easily using dplyr::mutate().

```
# Wrangle the dataset: Add in numeric variable conversion.

ds[72:93, 1:4] %>%
  set_names(c("period", "location", "export", "import")) %>%
  mutate(
    export = as.numeric(export),
    import = as.numeric(import)
  )

## # A tibble: 22 x 4
##     period  location export import
```

```
##          <chr>       <chr>  <dbl>  <dbl>
## 1 2001-02 Australia  99484  85235
## 2       <NA>  17 Ports  84597  83834
## 3 2002-03 Australia  93429  94947
## 4       <NA>  17 Ports  80170  93566
## 5 2003-04 Australia  89303  93467
## 6       <NA>  17 Ports  76163  92045
## 7 2004-05 Australia 106341 108923
....
```

Every second value of the `period` column is missing. This is typical of spreadsheet data where the table has an implied value of the cell above. We can use a simple indexing trick to replicate the `period` appropriately. The functions `base::seq()`, `base::rep()`, and `base::sort()` are quite handy here. We want to index the `period` column to repeat every second value so we want to effectively generate an index start from 1 then repeating the 1 followed by 3, 3, 5, 5, and so on up to 21. We combine this with `magrittr::extract2()` and `magrittr::extract()`.*

```
# Generate indicies that will be useful for indexing the data.

seq(1,21,2) %>% rep(2) %>% sort()

##  [1]  1  1  3  3  5  5  7  7  9  9 11 11 13 13 15 15 17 17 19
## [20] 19 21 21

# Confirm this achieves the desired outcome.

ds[72:93, 1:4] %>%
  set_names(c("period", "location", "export", "import")) %>%
  extract2("period") %>%
  extract(seq(1,21,2) %>% rep(2) %>% sort())

##  [1] "2001-02" "2001-02" "2002-03" "2002-03" "2003-04"
##  [6] "2003-04" "2004-05" "2004-05" "2005-06" "2005-06"
## [11] "2006-07" "2006-07" "2007-08" "2007-08" "2008-09"
....
```

Review the code above and be sure to understand how it has

*The function `extract()` is defined in multiple packages and we use the definition from magrittr here.

achieved the results and why this is useful. We will see that we can now dplyr::**mutate()** the dataset by replacing the original values of `period` with the appropriately repeated values of `period`. We add this step to the data wrangling sequence.

```
# Wrangle the dataset: Repair the period column.

ds[72:93, 1:4] %>%
  set_names(c("period", "location", "export", "import")) %>%
  mutate(
    export = as.numeric(export),
    import = as.numeric(import),
    period = period[seq(1, 21, 2) %>% rep(2) %>% sort()]
  )

## # A tibble: 22 x 4
##     period  location export import
##      <chr>     <chr>  <dbl>  <dbl>
##  1 2001-02 Australia  99484  85235
##  2 2001-02  17 Ports  84597  83834
##  3 2002-03 Australia  93429  94947
##  4 2002-03  17 Ports  80170  93566
##  5 2003-04 Australia  89303  93467
##  6 2003-04  17 Ports  76163  92045
....
```

We will now continue processing the dataset to reshape it into a form that suits the plot that we wish to generate. The required plot is a bar chart comparing the export and import totals for Australia and the largest 17 ports over a specific period of time. A common step is to "rotate" the dataset using tidyr::**gather()** as we do here. Prior to the tidyr::**gather()** the dataset has the four columns named `period`, `location`, `export`, and `import`. The result of the tidyr::**gather()** is a dataset with four columns named `period`, `location`, `type`, and `value`. The variable `type` has the values `Import` and `Export` and `value` is the numeric dollar value. This process is often referred to as *reshaping* the dataset.

```
# Wrangle the dataset: Reshape the datset.

ds[72:93, 1:4] %>%
  set_names(c("period", "location", "export", "import")) %>%
```

```
  mutate(
    export = as.numeric(export),
    import = as.numeric(import),
    period = period[seq(1, 21, 2) %>% rep(2) %>% sort()]
  ) %>%
  gather(type, value, -c(period, location))
```

```
## # A tibble: 44 x 4
##     period  location   type  value
##      <chr>     <chr>  <chr>  <dbl>
##  1 2001-02 Australia export  99484
##  2 2001-02  17 Ports export  84597
##  3 2002-03 Australia export  93429
##  4 2002-03  17 Ports export  80170
##  5 2003-04 Australia export  89303
##  6 2003-04  17 Ports export  76163
....
```

Our dataset is now looking rather tidy—it is in a form whereby we can readily pipe it into various plotting commands. In preparation for this we will define a colour scheme that is used consistently throughout the information sheet. We record the colours used here.* The approach is to generate a theme for the style used and to then add that theme into each plot. The choice of colours for plots is specified using `ggplot2::scale_fill_manual()`.

```
# Identify specific colors required for the organisaitonal style.

cols <- c('#F6A01A', # Primary Yellow
          '#0065A4', # Primary Blue
          '#455560', # Primary Accent Grey
          '#B2BB1E', # Secondary Green
          '#7581BF', # Secondary Purple
          '#BBB0A3', # Secondary Light Grey
          '#E31B23', # Secondary Red
          '#C1D2E8') # Variant Grey

# Create a ggplot2 theme using these colours.
```

*The actual colours used were provided by the author of the information sheet and developed by David Mitchell of the Bureau of Infrastructure, Transport and Regional Economics, based on the Department's themes, and used with permission.

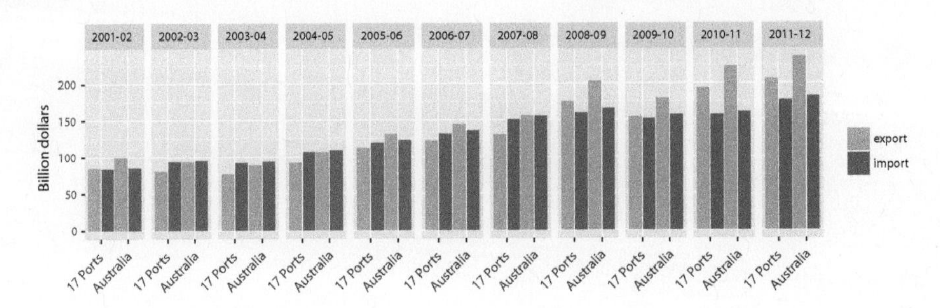

Figure 5.1: *Faceted dodged bar plot comparing the import and export dollar value across multiple years aggregated over the largest 17 ports and all Australian ports, respectively.*

```
theme_bitre <- scale_fill_manual(values=cols)
```

The following sequence of commands will wrangle the original data into a tidy dataset that is piped into `ggplot2::ggplot()` to generate a plot onto which various layers are added. The result can be seen in Figure 5.1.

```
ds[72:93, 1:4] %>%
  set_names(c("period", "location", "export", "import")) %>%
  mutate(
    export = as.numeric(export),
    import = as.numeric(import),
    period = period[seq(1, 21, 2) %>% rep(2) %>% sort()]
  ) %>%
  gather(type, value, -c(period, location)) %>%
  ggplot(aes(x=location, y=value/1000, fill=type)) +
  geom_bar(stat="identity", position=position_dodge(width=1)) +
  facet_grid(~period) +
  labs(y="Billion dollars", x="", fill="") +
  theme(axis.text.x=element_text(angle=45, hjust=1, size=10)) +
  theme_bitre
```

Obviously there is quite a lot happening in this sequence of commands. For the data wrangling we have worked through each step and added it into the pipeline one step at a time. Whist the resulting sequence looks complex, each step was quite straightforward. This is typical of the process of building our pipelines.

The plot command is also usually built in the same way whereby we add new layers or elements to plot one step at a time. The initial plot function itself identifies the aesthetics (`ggplot2::aes()`) having the `location` as the x-axis and the `value` divided by 1,000 as the y-axis. The `fill=` aesthetic specifies how to colour the plot. Here we choose to base the colour on the `type` (an import or an export).

We then added a `ggplot2::geom_bar()` to create a bar chart. The `ggplot2::facet_grid()` splits the plot into separate plots based on the `period`. We then change the label for the y-axis and remove labels on the x-axis and the legend. Finally we rotate the labels on the x-axis by 45° and then choose the colours for the fill.

Notice that this plot is different from that in the original information sheet. The bars are not stacked. Rather we use `position=` to specify they should be dodged. This allows the import pattern to be better understood from a common baseline rather than the varying baseline when the bars are stacked. Nonetheless the aim in the information sheet is to review the total value of the combined imports and exports and so the stacked bars are useful in that context.

Figure 3 of the information sheet is quite similar but this time the measure is the weight of the international sea trade. The dataset is similarly formatted in the spreadsheet and so we process it in the same way to generate Figure 5.2.

```
ds[96:117, 1:4] %>%
  set_names(c("period", "location", "export", "import")) %>%
  mutate(
    export = as.numeric(export),
    import = as.numeric(import),
    period = period[seq(1, 21, 2) %>% rep(2) %>% sort()]
  ) %>%
  gather(type, value, -c(period, location)) %>%
  ggplot(aes(x=location, y=value/1000, fill=type)) +
  geom_bar(stat="identity",position=position_dodge(width = 1)) +
  facet_grid(~period) +
  labs(y="Million tonnes", x="", fill="") +
  theme(axis.text.x=element_text(angle=45, hjust=1, size=10)) +
  theme_bitre
```

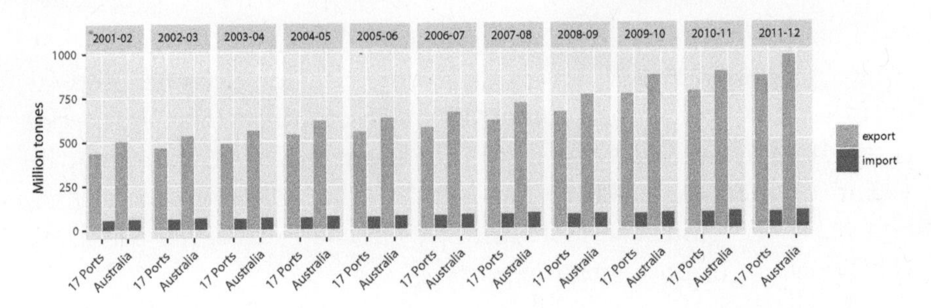

Figure 5.2: *Faceted dodged bar plot comparing the import and export total weight across multiple years aggregated over the largest 17 ports and all Australian ports, respectively.*

5.3 Scatter Plot: Throughput versus Annual Growth

Figure 4 of the information sheet presents a scatter plot charting the latest year's sea port throughput against the port's annual average growth rate. The first two tables in the spreadsheet are relevant to this plot. They record information about the 2011–12 throughput of each port and the port's 10-year annual average growth. Once again we need to wrangle this dataset into a form that is suitable for analyzing. We start with the first of the datasets identified as rows 3 and 4, with row 2 containing the port names.

```
# Confirm the table of interest from the dataset.

ds[2:4, 2:18]

## # A tibble: 3 x 17
##        X__2      X__3   X__4    X__5   X__6       X__7      X__8
##       <chr>     <chr>  <chr>   <chr>  <chr>      <chr>     <chr>
## 1 Adelaide Brisbane Burnie Dampier Darwin Devonport Fremantle
## 2     15.5        37      4     176     11          3        28
## 3      4.9         6    3.5     6.5     35          4       4.6
## # ... with 10 more variables: X__9 <chr>, X__10 <chr>,
## #   X__11 <chr>, X__12 <chr>, X__13 <chr>, X__14 <chr>,
## #   X__15 <chr>, X__16 <chr>, X__17 <chr>, X__18 <chr>
```

To tidy this dataset we will transpose it using `base::t()` and then ensure it remains a data frame by piping the result into `dplyr::data_frame()` and then `dplyr::tbl_df()`. Row names are excluded from the resulting data frame (`row.names=NULL`) and by setting `stringsAsFactors=FALSE` we keep strings as strings rather then converting them to factors.

```
# Wrangle the dataset: Transpose and retain as a dataset.

ds[2:4, 2:18] %>%
  t() %>%
  data.frame(row.names=NULL, stringsAsFactors=FALSE) %>%
  tbl_df()

## # A tibble: 17 x 3
##           X1    X2    X3
##        <chr> <chr> <chr>
## 1   Adelaide  15.5   4.9
## 2   Brisbane    37     6
## 3     Burnie     4   3.5
....
```

Meaningful column names can be added using the convenient support function `magrittr::set_names()`. The three columns are the port name, the throughput for 2011–12, and the 10-year annual average growth rate.

```
# Wrangle the dataset: Add renaming columns informatively.

ds[2:4, 2:18] %>%
  t() %>%
  data.frame(row.names=NULL, stringsAsFactors=FALSE) %>%
  tbl_df() %>%
  set_names(c("port", "weight", "rate"))

## # A tibble: 17 x 3
##             port weight  rate
##            <chr>  <chr> <chr>
## 1       Adelaide   15.5   4.9
## 2       Brisbane     37     6
## 3         Burnie      4   3.5
....
```

Using `dplyr::mutate()` we turn the numeric columns which are currently treated as character to numeric.

```
# Wrangle the dataset: Add in numeric variable conversion.

ds[2:4, 2:18] %>%
  t() %>%
  data.frame(row.names=NULL, stringsAsFactors=FALSE) %>%
  tbl_df() %>%
  set_names(c("port", "weight", "rate")) %>%
  mutate(
    weight = as.numeric(weight),
    rate   = as.numeric(rate)
  )

## # A tibble: 17 x 3
##            port weight  rate
##           <chr>  <dbl> <dbl>
##  1     Adelaide   15.5   4.9
##  2     Brisbane   37.0   6.0
##  3       Burnie    4.0   3.5
....
```

We then have a tidy dataset consisting of 17 observations (indexed by **port**) of two variables (**weight** and **rate**).

An extra feature of Figure 4 of the information sheet is that the ports are grouped into mixed and bulk ports. This information is contained in another table within the spreadsheet corresponding to rows 5 to 16 and columns 1 and 2 (with row 5 being the column names).

```
# Identify port types from appropriate data in the sheet.

ds[6, 1:2]

## # A tibble: 1 x 2
##     X__1  X__2
##    <chr> <chr>
## 1 Mixed  Bulk

ds[7:17, 1:2]

## # A tibble: 11 x 2
##            X__1         X__2
##           <chr>        <chr>
##  1   Melbourne      Dampier
```

```
## 2      Brisbane    Gladstone
## 3 Port Kembla    Hay Point
....
```

We can turn this into a data table listing the ports under their particular category.

```
# Construct a port type table.

ds[7:17, 1:2] %>%
  set_names(ds[6, 1:2])
```

```
## # A tibble: 11 x 2
##          Mixed           Bulk
##          <chr>          <chr>
##  1   Melbourne        Dampier
##  2    Brisbane      Gladstone
##  3 Port Kembla      Hay Point
##  4   Devonport      Newcastle
##  5      Sydney  Port Hedland
##  6     Geelong  Port Walcott
##  7    Adelaide           <NA>
##  8   Fremantle           <NA>
##  9      Darwin           <NA>
## 10      Burnie           <NA>
## 11  Townsville           <NA>
```

We can see that the Bulk column lists fewer ports than the Mixed column and thus we have missing values (`NA`) in the table. We will need to omit these from the final dataset using `stats::na.omit()` which we do below. First though we will use `tidyr::gather()` to tidy the dataset into two columns, one for the type of port and the other for the port name. This is a more useful data structure.

```
# Tidy the dataset into a more useful structure.

ds[7:17, 1:2] %>%
  set_names(ds[6, 1:2]) %>%
  gather(type, port) %>%
  na.omit()
```

```
## # A tibble: 17 x 2
```

```
##      type        port
##     <chr>       <chr>
##  1 Mixed   Melbourne
##  2 Mixed    Brisbane
##  3 Mixed Port Kembla
##  4 Mixed   Devonport
##  5 Mixed      Sydney
##  6 Mixed     Geelong
##  7 Mixed    Adelaide
##  8 Mixed   Fremantle
##  9 Mixed      Darwin
## 10 Mixed      Burnie
## 11 Mixed  Townsville
## 12  Bulk     Dampier
## 13  Bulk   Gladstone
## 14  Bulk   Hay Point
## 15  Bulk   Newcastle
## 16  Bulk Port Hedland
## 17  Bulk Port Walcott
....
```

We now have two source datasets that we can join into a single dataset using a dplyr::**left_join()**. This kind of operation will be familiar to database users.

```
# Wrangle the dataset: Join to port type.

ds[2:4, 2:18] %>%
  t() %>%
  data.frame(row.names=NULL, stringsAsFactors=FALSE) %>%
  tbl_df() %>%
  set_names(c("port", "weight", "rate")) %>%
  mutate(
    weight = as.numeric(weight),
    rate   = as.numeric(rate)
  ) %>%
  left_join(ds[7:17, 1:2] %>%
              set_names(ds[6, 1:2]) %>%
              gather(type, port) %>%
              na.omit(),
            by="port")

## # A tibble: 17 x 4
##             port weight  rate  type
```

```
##            <chr>  <dbl> <dbl> <chr>
## 1      Adelaide   15.5   4.9 Mixed
## 2      Brisbane   37.0   6.0 Mixed
## 3        Burnie    4.0   3.5 Mixed
....
```

We are now ready to reproduce Figure 4. With only a relatively small number of data points a scatter plot is a good choice as we can see in Figure 5.3. We have had to include some significant plotting features to achieve an effective presentation of the data here. The resulting code as a single block again is quite intimidating. However, stepping through the process, one element at a time, as we do in developing our code, each step is generally simple.

```
ds[2:4, 2:18] %>%
  t() %>%
  data.frame(row.names=NULL, stringsAsFactors=FALSE) %>%
  tbl_df() %>%
  set_names(c("port", "weight", "rate")) %>%
  mutate(
    weight = as.numeric(weight),
    rate   = as.numeric(rate)
  ) %>%
  left_join(ds[7:17, 1:2] %>%
              set_names(ds[6, 1:2]) %>%
              gather(type, port) %>%
              na.omit(),
            by="port") %>%
  mutate(type=factor(type, levels=c("Mixed", "Bulk"))) %>%
  filter(port != "Darwin") ->
tds

tds %>%
  ggplot(aes(x=weight, y=rate)) +
  geom_point(aes(colour=type, shape=type), size=4) +
  xlim(0, 300) + ylim(0, 13) +
  labs(shape="Port Type",
       colour="Port Type",
       x="Throughput 2011-12 (million tonnes)",
       y="Throughput average annual growth rate") +
  geom_text(data=filter(tds, type=="Bulk"),
            aes(label=port), vjust=2) +
  annotate("rect", xmin=0, xmax=45, ymin=3, ymax=6.5,
```

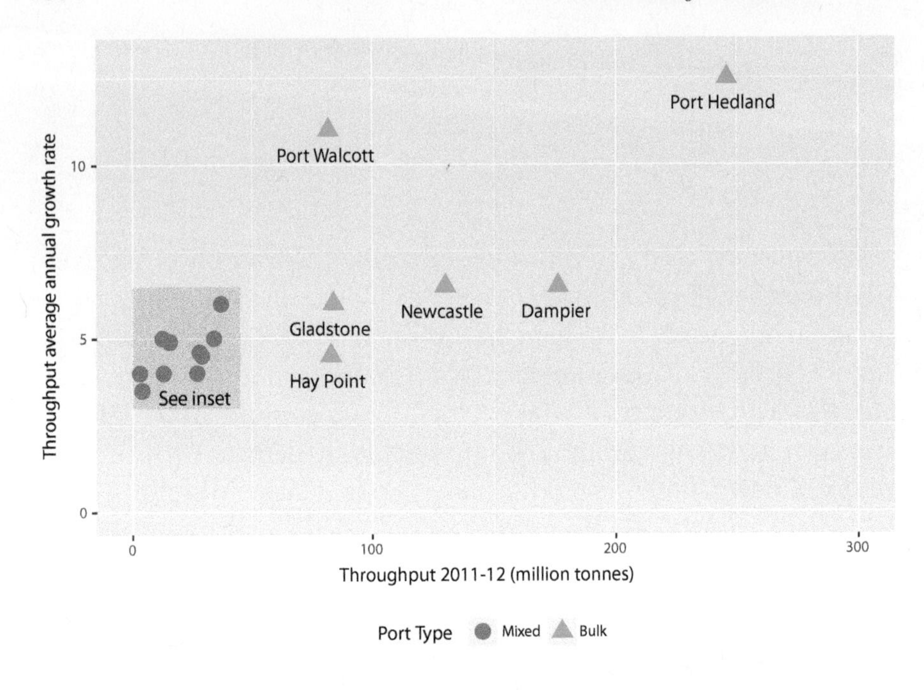

Figure 5.3: *Labelled scatter plot with inset.*

```
         alpha = .1) +
annotate("text", label="See inset", x=28, y=3.3, size=4) +
theme(legend.position="bottom")
```

After our initial gasp at seeing the complex sequence we can look at the detail—the process is made up of mostly simple steps. There are a couple of "tricks" here we should explain and we then again emphasise that the generation of this sequence of commands was an iterative process, building up one simple step at a time to achieve something quite complex.

The "tricks" were developed through several iterations of generating and reviewing the plot. The first was that we identified `Darwin` as an outlier which caused the plot to be poorly distributed. We decided to `dplyr::filter()` it out.

We also find the need to label just a subset of the points as the mixed ports are rather close together and labelling would be problematic. Instead we save the tidy dataset to a variable `tds` rather than piping it directly to `ggplot2::ggplot()`. In this way we might also reuse the dataset as required. We filter this dataset within ggplot2::geom_text() to label just a subset of the points.

Now we can take time to step through the ggplot2::ggplot() layers. First note how we map the x-axis and y-axis to specific variables. The layers then begin with ggplot2::geom_point() with the colour and shape of the points reflecting the type of the port: mixed or bulk. Shape is used in addition to colour as not everyone can distinguish colours as well. The size of the points is also increased.

The points within the scatter plot are labelled with the name of the port using ggplot2::geom_text(). As noted it was not practical to label the cluster of points representing the mixed ports so that region is shaded. We identify that an inset is available as a secondary plot shown in Figure 5.4.

For the secondary plot we again start with the saved tidy dataset tds, filter out points we do not wish to plot, overwriting tds so we can again use subsets of the dataset for labelling points differently.

```
above <- c("Townsville", "Fremantle")

tds %<>% filter(port != "Darwin" & type == "Mixed")

tds %>%
  ggplot(aes(x=weight, y=rate, label=port)) +
  geom_point(aes(colour=type, shape=type), size=4) +
  labs(shape="Port Type", colour="Port Type") +
  xlim(0, 40) + ylim(3, 6) +
  labs(x="Throughput 2011-12 (million tonnes)",
       y="Throughput average annual growth rate") +
  geom_text(data=filter(tds, !port%in%above), vjust= 2.0) +
  geom_text(data=filter(tds,  port%in%above), vjust=-1.0) +
  theme(legend.position="bottom")
```

The secondary plot focuses only on the mixed ports. At this scale there is plenty of room to label the points although we need to place a couple of the labels above the points rather than below. We also dplyr::filter() the original dataset to only include the Mixed ports, apart from Darwin. The assignment pipe magrittr::%<>% is used to overwrite the original temporary copy

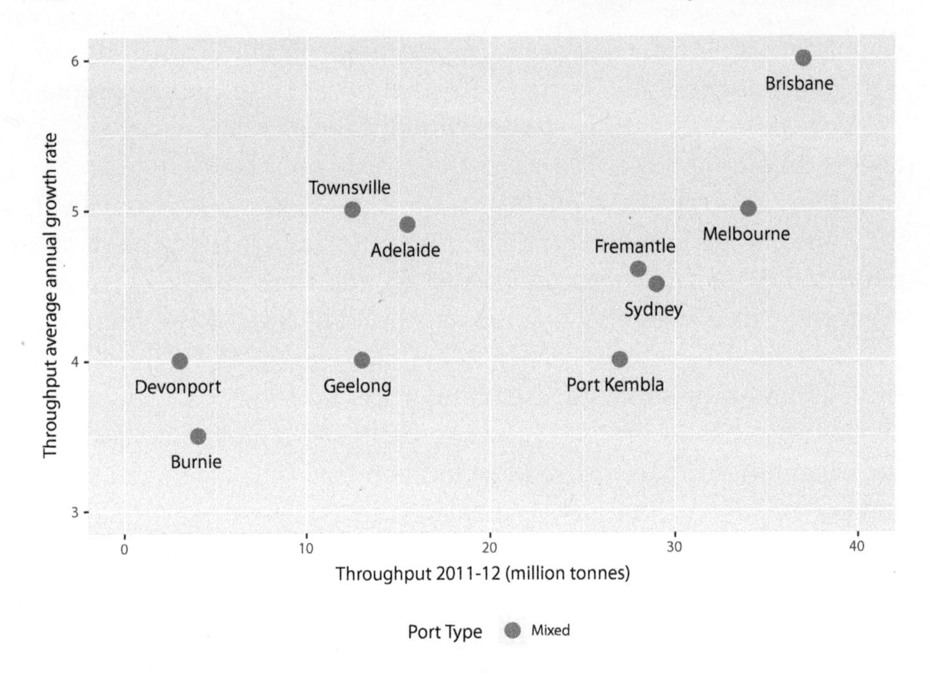

Figure 5.4: *Labelled scatter plot.*

of the dataset in `tds` which we no longer need. We would do so to refer to the whole dataset within `ggplot2::geom_text()` to simplify the coding. The remainder of the code is then similar to the previous scatter plot.

5.4 Combined Plots: Port Calls

Figure 5 in the information sheet combines two plots into one. This is advisable only where the data are closely related. The plot shows the number of ship calls to each port over 12 years. The plot is faceted by the port and so there are 17 separate plots. This leaves space in the bottom right for a secondary plot to show the average annual growth rate for each port. We might argue that this either overloads the plot or makes effective use of available space. We observe the latter here.

The calls data for the main plot is included in the spreadsheet as rows 20 to 36 and columns 1 to 13 with row 19 being the column names. We will build the calls dataset with the appropriate column names replacing the first with a more informative name.

```
# Wrangle the dataset: Name columns informatively.

ds[20:36, 1:13] %>%
  set_names(c("port", ds[19, 2:13]))

## # A tibble: 17 x 13
##              port `2001-02` `2002-03` `2003-04` `2004-05`
##             <chr>     <chr>     <chr>     <chr>     <chr>
##  1 Port Hedland       623       673       547       914
##  2    Melbourne      2628      2902      2935      3061
##  3    Newcastle      1452      1345      1382      1546
....
```

To tidy the dataset we tidyr::gather() the periods into a column and dplyr::mutate() the calls base::as.integer()s.

```
# Wrangle the dataset: Add dataset reshape and convert integer.

ds[20:36, 1:13] %>%
  set_names(c("port", ds[19, 2:13])) %>%
  gather(period, calls, -port) %>%
  mutate(calls=as.integer(calls))

## # A tibble: 204 x 3
##              port  period calls
##             <chr>   <chr> <int>
##  1 Port Hedland 2001-02   623
##  2    Melbourne 2001-02  2628
##  3    Newcastle 2001-02  1452
....
```

From the data we can also calculate the average annual growth using a formulation involving base::exp() and base::log(). This is required for the second plot.

```
# Wrangle the dataset: Add avg calculatation.

ds[20:36, 1:13] %>%
```

```
  set_names(c("port", ds[19, 2:13])) %>%
  select(port, 2, 13) %>%
  set_names(c('port', 'start', 'end')) %>%
  mutate(
    start = as.integer(start),
    end   = as.integer(end),
    avg   = 100*(exp(log(end/start)/11)-1)
  )

## # A tibble: 17 x 4
##              port start   end       avg
##             <chr> <int> <int>     <dbl>
##  1 Port Hedland   623  3920 18.200143
##  2    Melbourne  2628  3446  2.494151
##  3    Newcastle  1452  3273  7.668592
....
```

We now have the data required for the two plots. The two constituent plots are separately generated and saved in memory as the two R variables p1 and p2.

```
# Build the main faceted plot.

p1 <-
  ds[20:36, 1:13] %>%
  set_names(c("port", ds[19, 2:13])) %>%
  gather(period, calls, -port) %>%
  mutate(calls=as.integer(calls)) %>%
  ggplot(aes(x=period, y=calls)) +
  geom_bar(stat="identity", position="dodge", fill="#6AADD6") +
  facet_wrap(~port) +
  labs(x="", y="Number of Calls") +
  theme(axis.text.x=element_text(angle=90, hjust=1, size=8)) +
  scale_y_continuous(labels=comma)
```

```
# Generate the second plot.

p2 <-
  ds[20:36, 1:13] %>%
  set_names(c("port", ds[19, 2:13])) %>%
  select(port, 2, 13) %>%
  set_names(c('port', 'start', 'end')) %>%
```

```
  mutate(
    start = as.integer(start),
    end   = as.integer(end),
    avg   = 100*(exp(log(end/start)/11)-1)
  ) %>%
  ggplot(aes(x=port, y=avg)) +
  geom_bar(stat="identity",
           position="identity",
           fill="#6AADD6") +
  theme(axis.text.x=element_text(angle=45, hjust=1, size=8),
        axis.text.y=element_text(size=8),
        axis.title=element_text(size=10),
        plot.title=element_text(size=8),
        plot.background = element_blank()) +
  labs(x="",
       y="Per cent",
       title="Average Annual Growth, 2001-02 to 2012-13")
```

The use of `grid::viewport()` now allows us to insert one plot over another resulting in Figure 5.5.

```
# Combine the plots into a single plot.

print(p1)
print(p2, vp=viewport(x=0.72, y=0.13, height=0.28, width=0.54))
```

5.5 Further Plots

The information sheet goes on to produce a variety of other plots to support the narrative presented. Below we illustrate the coding that can be used to generate a number of these plots. As noted earlier, the narrative or the story that is told and supported by the analysis is left to the information sheet itself (BITRE, 2014).

For each of the plots the programming code presented here is complete. The overall data wrangling and visualisation code may appear quite complex for each but it is important to yet again emphasise that the final code is developed step-by-step. Only the final sequence of code is presented to wrangle the data and to

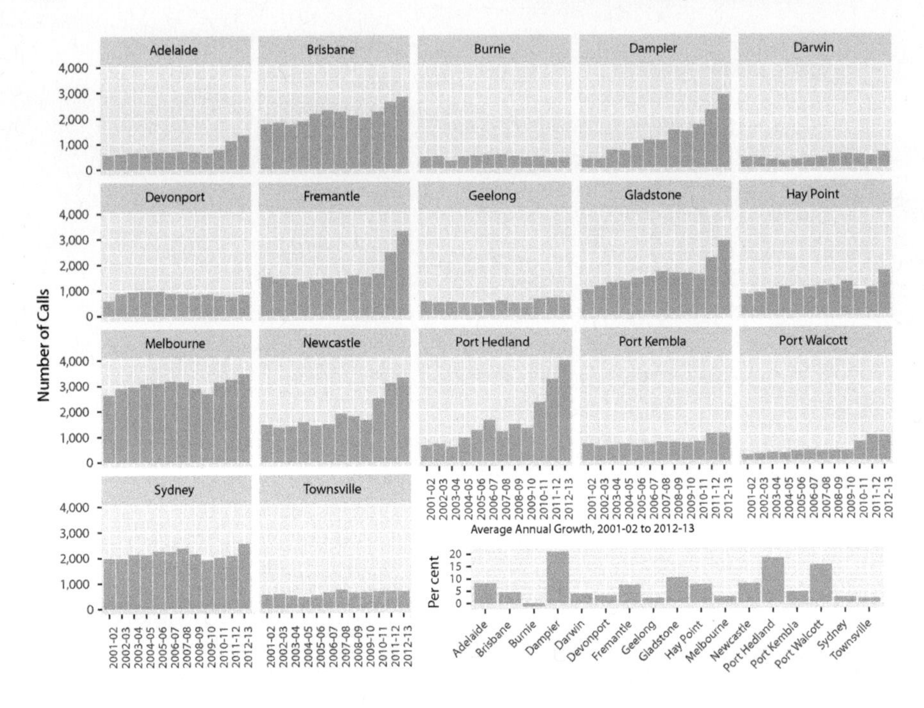

Figure 5.5: *Faceted bar plot with embedded bar plot.*

produce the plot. Taken one step at a time we can often build up to the full plot quite simply and are able to understand the simplicity of each step as we build up to quite complex processing.

The next three figures from the information sheet are sideways bar charts and we'll reproduce one of them here (Figure 7a). As always we begin by wrangling the data to produce a tidy dataset. We plot the percentages in each of the occupation groups. The data can be found in rows 48 to 56 over two columns. We wrangle the data as usual and generate the plot using ggplot2::ggplot() shown in Figure 5.6.

```
ds[48:56, 1:2] %>%
  set_names(c("occupation", "percent")) %>%
  mutate(
    percent    = as.numeric(percent),
    occupation = factor(occupation,
                        levels=occupation[order(percent)])
  ) %>%
```

Figure 5.6: *Horizontal bar chart.*

```
  ggplot(aes(x=occupation, y=percent)) +
  geom_bar(stat="identity", fill="#6AADD6", width=0.8) +
  theme(axis.title.x=element_text(size=10)) +
  labs(x="", y="Per cent") +
  coord_flip()
```

Notice how we ensure the order of the occupations within the plot to correspond to the percentage representation. The second `dplyr::mutate()` transforms `occupation` into a `factor` with levels having the order based on the `percent`age.

Another interesting aspect of the plot is the use of the convenient `ggplot2::coord_flip()` which rotates the plot. This generates the sideways bars.

Figure 8 of the information sheet is a visually interesting horizontal bar chart with annotations. We take this in two steps because the labelling requires further processing of the data. The wrangled dataset is saved as the variable `tds`.

```
tds <-
  ds[39:40, 2:9] %>%
  set_names(ds[38, 2:9]) %>%
  mutate(type=c("Mixed Ports", "Bulk Ports")) %>%
  gather(occupation, percent, -type) %>%
  mutate(
    percent    = as.numeric(percent),
    occupation = factor(occupation,
```

```
         levels=c("Managers",
                  "Professionals",
                  "Technicians and Trades Workers",
                  "Community and Personal Service Workers",
                  "Clerical and Administrative Workers",
                  "Sales Workers",
                  "Machinery Operators and Drivers",
                  "Labourers"))
  ) %T>%
  print()

## # A tibble: 16 x 3
##          type                           occupation percent
##         <chr>                               <fctr>   <dbl>
##  1 Mixed Ports                            Managers    12.2
##  2  Bulk Ports                            Managers     8.6
##  3 Mixed Ports                       Professionals    15.1
....
```

To be able to annotate the plot with the percentages we need to identify the correct positions for each one. Here we construct a data frame which has the locations of the labels (x and y) together with the labels themselves (v).

```
mv <-
  tds %>%
  filter(type=="Mixed Ports") %>%
  extract2("percent") %>%
  rev()

my <- (mv/2) + c(0, head(cumsum(mv), -1))

bv <-
  tds %>%
  filter(type=="Bulk Ports") %>%
  extract2("percent") %>%
  rev()

by <- (bv/2) + c(0, head(cumsum(bv), -1))

lbls <-
  data.frame(x=c(rep(1, length(mv)), rep(2, length(bv))),
             y=c(by, my),
             v=round(c(bv, mv))) %T>%
```

```
  print()

##    x     y  v
## 1  1  6.85 14
## 2  1 21.15 15
## 3  1 29.80  2
## 4  1 37.10 12
## 5  1 45.40  4
....
```

The data to generate the plot is now ready. The plot is a simple bar chart with stacked bars (the default) with each bar labelled by its size and coloured according to the theme. We flip the bar chart to have it horizontal and reverse the y values to match the legend.

```
tds %>%
  ggplot(aes(x=type, y=percent, fill=occupation)) +
  geom_bar(stat="identity", width=0.5) +
  labs(x="", y="Per cent", fill="") +
  annotate("text",
           x=lbls$x,
           y=lbls$y,
           label=lbls$v,
           colour="white") +
  coord_flip() +
  scale_y_reverse() +
  theme_bitre
```

Figure 9 of the information sheet compares the proporation of employees working full time or long hours at the different types of ports (bulk versus mixed versus combined). Figure 5.8 reproduces the plot.

```
ds[43:45, 1:3] %>%
  set_names(c("type", ds[42, 2:3])) %>%
  gather(var, count, -type) %>%
  mutate(
    count = as.integer(count),
    type  = factor(type,
                   levels=c("Bulk", "Mixed", "Australia"))
  ) %T>%
```

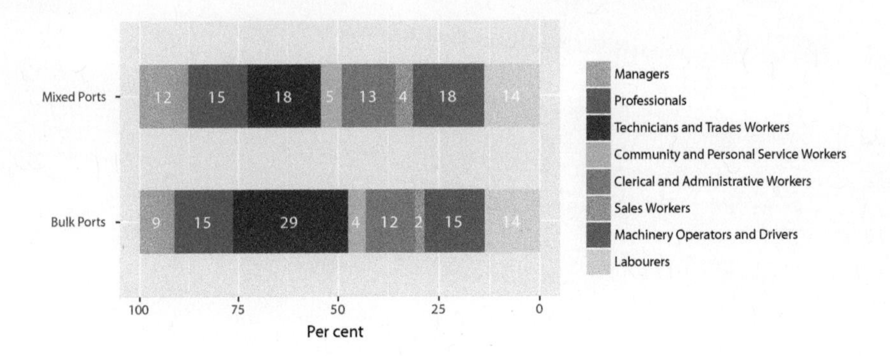

Figure 5.7: *Horizontal bar chart with multiple stacks.*

```
  print() ->
tds

## # A tibble: 6 x 3
##        type                              var count
##      <fctr>                            <chr> <int>
## 1      Bulk      Share of full time workers    88
## 2     Mixed      Share of full time workers    83
## 3 Australia      Share of full time workers    68
....

lbls <- data.frame(x=c(.7, 1, 1.3, 1.7, 2, 2.3),
                   y=tds$count-3,
                   lbl=round(tds$count))

tds %>%
  ggplot(aes(x=var, y=count)) +
  geom_bar(stat="identity", position="dodge", aes(fill=type)) +
  labs(x="", y="Per cent", fill="") + ylim(0, 100) +
  geom_text(data=lbls,
            aes(x=x, y=y, label=lbl),
            colour="white") +
  theme_bitre
```

Throughout this chapter we have presented significant sample code. We have reinforced that the development of the final code we see here for each plot is an iterative process where step-by-

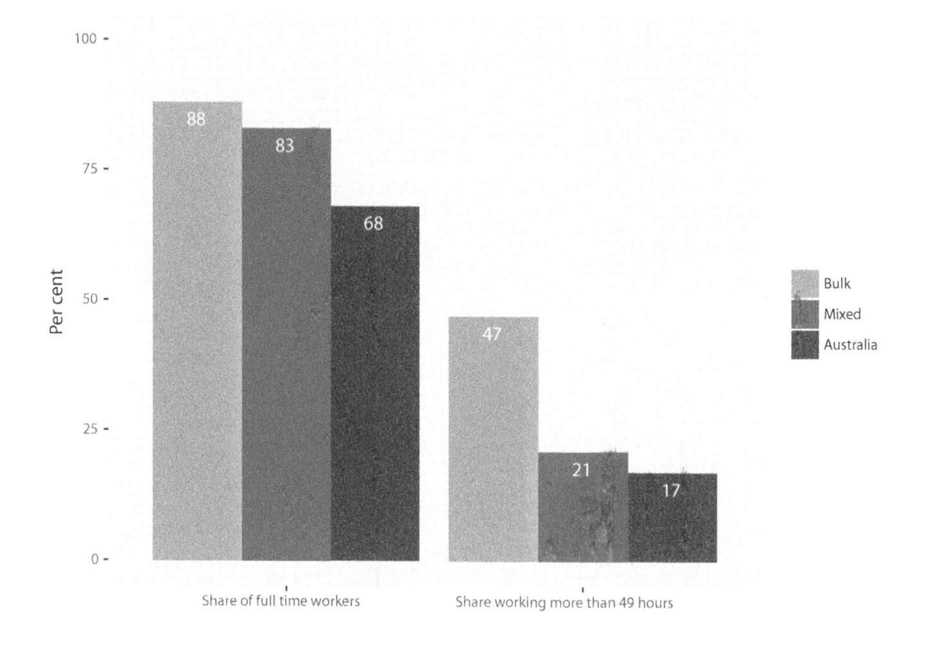

Figure 5.8: *Simple bar chart with dodged and labelled bars.*

step we have added processing of the dataset to achieve one action at a time. Combined we end up with quite complex operations to produce powerful plots that then support the narrative of the report.

5.6 Exercises

Exercise 5.1 Simple Bar Chart

Figure 7a of the BITRE report is a simple horizontal bar chart and was reproduced above as Figure 5.6. Using that code as a template reproduce Figures 6 and 7b of the report.

Exercise 5.2 Stacked Bar Chart

Figure 8a of the BITRE report is a stacked bar chart and was reproduced above as Figure 5.7. Using that code as a template reproduce Figures 8b and 11 of the report.

Exercise 5.3 Scatter Plots

Figure 10 of the BITRE report contains two scatter plots similar to that of Figure 4 of the report which we reproduced above as Figure 5.3. Using that code as a template reproduce Figure 10.

6

Case Study: Web Analytics

Open data is a focus for many governments who have identified the value to society and business in making data available for anyone to access and analyse. The Australian government, like many others, makes open data available through the data.gov.au repository. The Australian Taxation Office publishes a variety of open data on the repository including web log data. For our purposes we will use this web log data as another use case for a data scientist.

Many data repositories today provide application programmer interfaces (**APIs**) through which to access their data holdings. This allows us to more readily discover datasets and to readily load them into R, for example, for analysis. The Comprehensive Knowledge Archive Network (known as **CKAN**) provides a standard API to access data. This API is supported for accessing datasets stored on data.gov.au and many other similar repositories.

In this chapter we illustrate the process of accessing data from a CKAN node, loading it into R and then performing a data analysis on the dataset using the tools we have presented in the previous chapters.

Packages used in this chapter include ckanr (Chamberlain, 2015), dplyr (Wickham *et al.*, 2017a), ggplot2 (Wickham and Chang, 2016), magrittr (Bache and Wickham, 2014), rattle (Williams, 2017), readr (Wickham *et al.*, 2017b), scales (Wickham, 2016), stringi (Gagolewski *et al.*, 2017), stringr (Wickham, 2017a), tidyr (Wickham, 2017b) and xtable (Dahl, 2016)

```
# Load required packages from local library into R.

library(ckanr)     # Access data from CKAN.
library(dplyr)     # Wrangling: group_by().
library(ggplot2)   # Visualise data.
library(magrittr)  # Pipelines for data processing: %>% %T>% %<>%.
```

```
library(rattle)    # Wrangling: normVarNames().
library(readr)     # Modern data reader.
library(scales)    # Include commas in numbers in plots.
library(stringi)   # The string concat operator: %s+%.
library(stringr)   # String manpiulation: str_split().
library(tidyr)     # Tidy the dataset: gather().
library(xtable)    # Format R data frames as LaTeX tables.
```

6.1 Sourcing Data from CKAN

The Australian Taxation Office has made available for public access amongst many other datasets a web log dataset. This dataset was originally made available for the Australian government's GovHack 2014 contest. The dataset can be downloaded from the ATO Web Analytics section of the data.gov.au repository using any browser. Since the repository supports the CKAN API we can use ckanr to program our navigation through to the dataset of interest and then load that dataset into R.

There are many repositories that support the CKAN API and we can list some of them using `ckanr::servers()`.

```
# List some of the known servers.

servers()

##   [1] "http://catalog.data.gov"
##   [2] "http://africaopendata.org"
##   [3] "http://annuario.comune.fi.it"
##   [4] "http://bermuda.io"
##   [5] "http://catalogue.data.gov.bc.ca"
##   [6] "http://catalogue.datalocale.fr"
....

# Count the number of servers.

servers() %>% length()

## [1] 124
```

Since we are interested in the Australian government's CKAN repository rather than the default provided by ckanr we need to set up its URL as the default.

```
# Current default server.

get_default_url()

## [1] "http://data.techno-science.ca/"

# Change the default server.

ckanr_setup(url="http://data.gov.au")

# Check the new default server.

get_default_url()

## [1] "http://data.gov.au"
```

A variety of metadata is available about each CKAN repository. This includes the various organisations that contribute to the repository. For the Australian government repository some 521 organisations are listed. For each organisation the metadata provided by the repository includes a description of the organisation and its logo, for example. We can list what metadata is available using `base::names()`.

```
# Store the list of organisations.

orgs <- organization_list(as="table")

# Report the number of organisations listed.

nrow(orgs)

## [1] 521

# Identify metadata available.

names(orgs)
```

```
##  [1] "display_name"       "description"
##  [3] "image_display_url" "package_count"
##  [5] "created"            "name"
##  [7] "is_organization"    "state"
##  [9] "image_url"          "type"
## [11] "title"              "revision_id"
## [13] "num_followers"      "id"
## [15] "approval_status"
```

Notice that the variable names here are already in our preferred style and so we have no need for our usual step of normalising them with rattle::**normVarNames()**.

The actual dataset we will use was provided by the Australian Taxation Office so we can find the ID of the organisation which we will use as a key to access the datasets provided by that organisation. We will dplyr::**filter()** and tidyr::**extract()** its internal ID. In the process we will list the count of the number of *packages* provided by the organisation. This is base::**print()**'d within a tee pipe so that after printing, the original data is piped on so as to magrittr::**extract()** the ID.

```
# Extract and store the organisation of interest's identifier.

orgs %>%
  filter(title=="Australian Taxation Office") %>%
  extract(c("id", "package_count")) %T>%
  print() %>%
  extract("id") ->
ato_id

##                                      id package_count
## 1 90d1f157-c01f-4589-93bf-600dee01996e            14
```

The organisation ID will be used to search for packages (which collect together possibly several datasets) from the organisation using ckanr::**package_search()**. Each of the packages will be listed under the results which we store as `ato_pkgs` in the following code block. We also report on the number of packages found and the title of the packages.

```
# Pattern is key:string

pattern  <- ("owner_org:" %s+% ato_id) %T>% print()

## [1] "owner_org:90d1f157-c01f-4589-93bf-600dee01996e"

# Find all packages matching the pattern.

ato_pkgs <- package_search(pattern, rows=100)$results

# Report the number of packages found.

length(ato_pkgs)

## [1] 14

# List the title of each package found.

sapply(ato_pkgs, extract2, 'title')

##  [1] "Taxation Statistics 2014-15"
##  [2] "Taxation Statistics 2011-12"
##  [3] "Voluntary Tax Transparency Code"
##  [4] "Taxation statistics - Individual sample files"
##  [5] "Ad-hoc requested data"
##  [6] "Taxation Statistics 2012-13"
##  [7] "Taxation Statistics 2013-14"
##  [8] "Taxation Statstics 1994-95 to 2008-09"
##  [9] "Taxation Statistics 2010-11"
## [10] "Taxation Statistics 2009-10"
## [11] "Corporate Tax Transparency"
## [12] "GovHack2016"
## [13] "Cumulative Total Tax Returns Received"
## [14] "ATO Web Analytics July 2013 to April 2014"
```

We can see the dataset we are interested in there as `ATO Web Analytics July 2013 to April 2014`. We'll extract just that package.

```
# Pattern is key:string

pattern     <- "title:ato web analytics"
```

```
# Search for specific package.

ato_web_pkg <- package_search(pattern)$results[[1]] %T>% print()

## <CKAN Package> 9b57d00a-da75-4db9-8899-6537dd60eeba
##   Title: ATO Web Analytics July 2013 to April 2014
##   Creator/Modified: 2014-05-14T02:09:38.206001 / 2014-05-1...
##   Resources (up to 5): Browser by month and traffic source...
##   Tags (up to 5): ATO, tax stats, web analytics, website s...
##   Groups (up to 5):

# Save the package identifier.

pid <- ato_web_pkg$id %T>% print()

## [1] "9b57d00a-da75-4db9-8899-6537dd60eeba"
```

There are 7 resources—a resource corresponds to a specific dataset.

```
ato_web_pkg$num_resources

## [1] 7
```

The actual list of datasets is included within the resources metadata of the package and we can list their names.

```
ato_web_pkg$resources %>% sapply(extract2, 'name')

## [1] "Browser by month and traffic source - July 2013 to Ap...
## [2] "Entry pages by month and traffic source - July 2013 t...
## [3] "Entry referrers by month and traffic source - July 20...
## [4] "Exit pages by month and traffic source - July 2013 to...
## [5] "Local keywords (top 100) by month and traffic source ...
## [6] "Operating System (platform) by month and traffic sour...
....
```

The first dataset listed identifies itself as browser data and is the dataset we will consider first.

6.2 Browser Data

Commonly datasets (or resources) on a CKAN repository like the browser dataset have metadata available that describes aspects of the dataset. We can view the type of metadata as the `base::names()` of the resource data structure. Recalling that the browser dataset is the first resource in the ATO Web Analytics package we access the resource appropriately.

```
# Save the resource structure for the web analytics dataset.

bwres <- ato_web_pkg$resources[[1]]

# Available metadata.

names(bwres)

##  [1] "cache_last_updated"     "package_id"
##  [3] "webstore_last_updated" "autoupdate"
##  [5] "datastore_active"       "id"
##  [7] "size"                   "state"
##  [9] "hash"                   "description"
## [11] "format"                 "last_modified"
## [13] "url_type"               "mimetype"
## [15] "cache_url"              "name"
## [17] "created"                "url"
## [19] "webstore_url"           "mimetype_inner"
## [21] "position"               "revision_id"
## [23] "resource_type"
```

Notice that there is a description field which we can access as `bwres$description` which describes this dataset as: *Browsers used to access the ATO website from July 2013 to April 2014. Data is broken down by month and by traffic source (internal or external).*

We are particularly interested in how to access the dataset and the format of the dataset. The `url` metadata specifies the location of the file containing the dataset. The `format` indicates that a ZIP archive is used to contain a CSV file. The CSV file contains the actual data. The base name of the CSV file that is contained within the ZIP archive is provided by the **name** metadata.

```
bwres$url

## [1] "http://data.gov.au/dataset/9b57d00a-da75-4db9-8899-65...

bwres$format

## [1] "ZIP (CSV)"

bwres$name

## [1] "Browser by month and traffic source - July 2013 to Ap...
```

Ingestion

We are now in a position to download the dataset into R so as to begin our analysis. The next code block does this by recording the URL and the CSV filename and creating a temporary filename to store the downloaded ZIP file. Using `utils::download.file()` we obtain a local copy of the file. When executing this interactively we notice that the progress of the download will be displayed. We then `base::unz()` the locally saved file and read it as CSV using `readr::read_csv()` to save it into the R variable `browsers`. The `base::unlink()` does the housekeeping to remove the downloaded ZIP file.

```
dspath   <- bwres$url
csvname  <- bwres$name %s+% ".csv"
temp     <- tempfile(fileext=".zip")

download.file(dspath, temp)
browsers <- unz(temp, csvname) %>% read_csv()
unlink(temp)
```

Preparation

Using our template from Chapter 3 we take a copy of the dataset into the template variable `ds`. We also have our first `dplyr::glimpse()` of the dataset.

```
dsname <- "browsers"
ds     <- get(dsname)
glimpse(ds)
```

```
## Observations: 1,357
## Variables: 5
## $ Browser        <chr> "Chrome", "Chrome", "Microsoft Inte...
## $ Month          <chr> "Jul-13", "Jul-13", "Jul-13", "Jul-...
## $ Traffic Source <chr> "External", "Internal", "External",...
## $ Views          <int> 7765921, 454, 6773492, 6067298, 507...
## $ Visits         <int> 691120, 110, 557509, 519093, 455815...
```

We observe that the dataset consists of 1,357 observations of 5 variables. Immediately we notice that the variable names are not in our preferred normalised form. As usual we convert them appropriately so that there is no ambiguity.

```
names(ds) %<>% normVarNames() %T>% print()
```

```
## [1] "browser"        "month"          "traffic_source"
## [4] "views"          "visits"
```

Optionally we decide to simplify one of the variable names noting that it is going to be simpler for us to refer to `traffic_source` as `source`.

```
names(ds)[3] <- "source"
```

Taking another `dplyr::glimpse()` of the dataset we can see some opportunity for structural modifications.

```
glimpse(ds)
```

```
## Observations: 1,357
## Variables: 5
## $ browser <chr> "Chrome", "Chrome", "Microsoft Internet Ex...
```

```
## $ month   <chr> "Jul-13", "Jul-13", "Jul-13", "Jul-13", "J...
## $ source  <chr> "External", "Internal", "External", "Exter...
## $ views   <int> 7765921, 454, 6773492, 6067298, 5078805, 6...
## $ visits  <int> 691120, 110, 557509, 519093, 455815, 69007...
```

The first three variables are generic character (`chr`) data types. The variable `month` is clearly a date and we could transform it into a suitable date format but for our simpler purposes we'll retain it as a factor. It has just 10 unique values and so will have this many levels as a factor. We retain the ordering of the months in the dataset to order the levels for the factor as per chance the order in the dataset is sequentially correct.

```
length(unique(ds$month))

## [1] 10

unique(ds$month)

##  [1] "Jul-13" "Aug-13" "Sep-13" "Oct-13" "Nov-13" "Dec-13"
##  [7] "Jan-14" "Feb-14" "Mar-14" "Apr-14"

ds %<>% mutate(month=factor(month, levels=unique(ds$month)))
```

We will similarly map `source` into a factor noting that it has just 2 values and it is okay to retain an alphabetic ordering to the levels.

```
length(unique(ds$source))

## [1] 2

unique(ds$source)

## [1] "External" "Internal"

ds %<>% mutate(source=factor(source))
```

The `browser` is also a candidate to be represented as a factor. We notice that the variable has 406 `base::unique()` values.

That's quite a lot and there is plenty of opportunity to tidy up this variable.

We start by recognising that multiple versions of the Microsoft Internet Explorer are reported and we might take the opportunity to combine these into a single entry. Reviewing the full list also suggests opportunities to combine others into single groups, such as the suite of Nokia identified browsers. A simple heuristic might suffice for our purposes so that stripping all but the first word from the browser string will be our starting point.

```
ds %<>% mutate(browser=
                  str_split(ds$browser, "\\W") %>%
                  sapply(extract, 1))
```

That brings us down to just 168 browsers which is still quite a few. We consider other opportunities to reduce the noise in relation to the analysis we are interested in. It is important to also note that we are removing potentially important information from our dataset and we need to do so with an understanding of the analysis we want to undertake and with care.

Many of the reported browsers appear infrequently and so we will aggregate them into a new group called *Others*. We need to sum the visits per browser over the whole dataset and using a threshold any browser not frequent enough will be renamed as *Other*.

```
# Threshold below which a browser is considered as Other.

visits.threshold <- 1e4

# Determine the list of Other browsers.

ds %>%
  group_by(browser) %>%
  summarise(visits=sum(visits)) %>%
  filter(visits < visits.threshold) %>%
  extract2('browser') ->
other

# Helper function.
```

```
remap.other <- function(x) if (x %in% other) "Other" else x

# Remap the browsers noting the opportunity to collapse
# repeated browser/month/source entries. Notice we need to
# remove the grouping to facilitate fuhter processing.

ds %<>%
  mutate(browser=sapply(browser, remap.other)) %>%
  group_by(browser, month, source) %>%
  summarise(views=sum(views), visits=sum(visits)) %>%
  ungroup()
```

Having transformed the character strings that record the browser type we are now ready to convert the data into a factor.

```
# Check the number of unique browsers we now have.

length(unique(ds$browser))

## [1] 9

# Record the browser names ordered by their
# frequency of visits.

ds %>%
  group_by(browser) %>%
  summarise(visits=sum(visits)) %>%
  arrange(visits) %>%
  extract2("browser") %>%
  as.character() %T>%
  print() ->
blvls

## [1] "Other"     "Netscape"  "Opera"     "Mozilla"   "Safar...
## [6] "Firefox"   "Mobile"    "Chrome"    "Microsoft"

# Convert browser into a factor with the sorted levels.

ds %<>% mutate(browser=factor(browser, levels=blvls))
```

Once again we `dplyr::glimpse()` the dataset to review our progress.

```
glimpse(ds)

## Observations: 122
## Variables: 5
## $ browser <fctr> Chrome, Chrome, Chrome, Chrome, Chrome, C...
## $ month   <fctr> Jul-13, Jul-13, Aug-13, Aug-13, Sep-13, S...
## $ source  <fctr> External, Internal, External, Internal, E...
## $ views   <int> 7765921, 454, 5625394, 296, 3563878, 747, ...
## $ visits  <int> 691120, 110, 665842, 78, 504094, 52, 75150...
```

We now have a dataset that has been sourced from a CKAN server. We have ingested that data into R as a data frame. From a review of the data and indeed from the name of the dataset and the description from the metadata we confirm that the observations range over the period from July 2013 to April 2014 by month and by the traffic source which is either internal or external. This is in line with what we were expecting for this dataset. We are ready for our analysis.

Analysis

We begin our analysis with an exploration of the internal versus external profiles of browser usage. First we `dplyr::group_by()` the values of the `source` and then `dplyr::summarise()` the dataset to compare the external and internal visits.

```
ds %>%
  group_by(source) %>%
  summarise(total=sum(visits)) %T>%
  print() ->
freq

## # A tibble: 2 x 2
##     source    total
##     <fctr>    <int>
## 1 External 33946553
## 2 Internal  1103712

round(100*freq$total[2]/sum(freq$total))

## [1] 3
```

Internal visits account for just 3% of all visits. This is not an unexpected ratio as internal visits come from the employees of the organisation which is a significantly smaller population than the external population.

For a more professional presentation of the results as we would produce for our actual report we would use the output from xtable::**xtable()** and leave it to LaTeX* to format it appropriately. Here we also have the titles of the columns bold and provide a caption for the table. The results can be seen in Table 6.1.

```
bold <- function(x)
{
  ifelse(x == "Visits",
         "\\multicolumn{1}{c}{\\textbf{Visits}}",
         '{\\textbf{' %s+% x %s+% '}}')
}
caption <-
  "External versus internal visits " %s+%
  "to the ATO web site together with the " %s+%
  "number of unique browsers identified."
short <-
  "External versus internal visits."

ds %>%
  group_by(source) %>%
  summarise(
    total    = sum(visits),
    browsers = length(unique(browser))
  ) %>%
  set_names(c("Source", "Visits", "Browsers")) %>%
  xtable(
    caption = c(caption, short),
    label   = "tab:atoweb:browser_visits"
  ) %>%
  print(include.rownames=FALSE,
        format.args=list(big.mark=","),
        sanitize.colnames.function=bold,
        table.placement="t",
        caption.placement="top",
        booktabs=TRUE)
```

*LaTeX is the free and open source typesetting system used for this book. See Chapter 10 for details.

Table 6.1: *External versus internal visits to the ATO web site together with the number of unique browsers identified.*

Source	Visits	Browsers
External	33,946,553	9
Internal	1,103,712	5

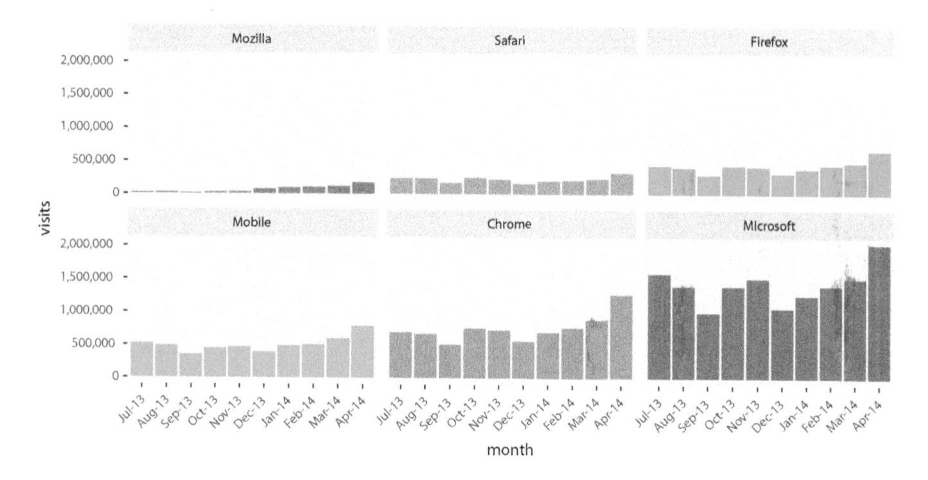

Figure 6.1: *Month by month external browser visits for the most popular browsers.*

An interesting analysis might be around the profiles of browsers used and their changing patterns of usage over time. Given the large number of browsers reported in the data we will ignore those less frequently represented browsers. We will also look at the external and internal visitors separately. Finally, a visual presentation can have more impact in our reports and so we will generate a plot as the output of this analysis. The result is Figure 6.1.

```
ds %>%
  filter(browser %in% levels(browser)[4:9]) %>%
  filter(source == "External") %>%
  ggplot(aes(month, visits, fill=browser)) +
  geom_bar(stat="identity") +
  facet_wrap(~browser) +
  scale_y_continuous(labels=comma) +
  theme(axis.text.x=element_text(angle=45, hjust=1)) +
```

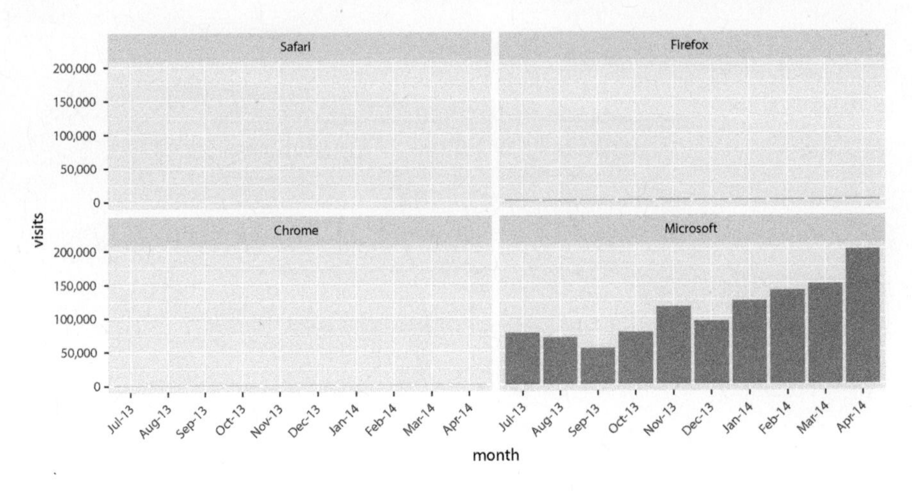

Figure 6.2: *Month by month internal browser visits for the most popular browsers.*

```
theme(legend.position="none")
```

Reviewing the plot provides insight into the behaviour of external users accessing the web site. We might postulate a seasonal interest in access to the site. The monthly patterns across the browsers are vaguely similar from month to month except for the Mozilla browser which is not present early on and increases in presence from December 2013. We might seek to gain some understanding of those observations and then test specific seasonal observations statistically.

In comparing these patterns with the usage profiles for internal users we see the plot of Figure 6.2. The two plots demonstrate a clear behavioural difference between external and internal users. There are considerably less visits and interestingly we observe predominantly Microsoft only browser visits. This reflects the typical organisational standard operating environments that dictate the usage of specific browsers. Otherwise we would begin to delve into the data collection itself to determine there is no data collection issue with this dataset.

Table 6.2: *External versus internal visits to the ATO web site by specific web browsers.*

Browser	External	Internal
Other	14,085	2
Netscape	29,847	
Opera	60,554	
Mozilla	634,484	
Safari	2,282,737	1
Firefox	4,296,359	12,793
Mobile	5,016,803	
Chrome	7,508,320	792
Microsoft	14,103,364	1,090,124

```
ds %>%
  filter(browser %in% levels(browser)[4:9]) %>%
  filter(source == "Internal") %>%
  ggplot(aes(month, visits, fill=browser)) +
  geom_bar(stat="identity") +
  facet_wrap(~browser) +
  scale_y_continuous(labels=comma) +
  theme(axis.text.x=element_text(angle=45, hjust=1)) +
  theme(legend.position="none")
```

All such observations need to be supported through a review of the actual data. Whilst the visualisations are appealing, presentations of the actual data remain fundamental to telling our narrative. A table of browser visits by source will confirm that the Microsoft browsers overwhelm the usage of other browsers internally within this organisation. There are only minor visits internally using Firefox. Table 6.2 is generated through the following code block.

```
caption <-
  "External versus internal visits " %s+%
  "to the ATO web site by specific " %s+%
  "web browsers."
short <-
  "External versus internal browsers."
```

```
ds %>%
  group_by(browser, source) %>%
  summarise(total=sum(visits)) %>%
  spread(source, total) %>%
  set_names(c("Browser", "External", "Internal")) %>%
  xtable(
    caption = c(caption, short),
    label   = "tab:atoweb:browser_split_source"
  ) %>%
  print(include.rownames=FALSE,
        format.args=list(big.mark=","),
        sanitize.colnames.function=bold,
        table.placement="t",
        caption.placement="top",
        booktabs=TRUE)
```

6.3 Entry Pages

Another dataset contained within the package records the starting (entry) pages for the ATO web site. As we did for the browser dataset we will load the entry dataset using the CKAN API. This is the second dataset in the package.

```
ato_web_pkg$resources[[2]]$name

## [1] "Entry pages by month and traffic source - July 2013 t...
```

The URL of the dataset can be obtained from the resource meta data from which we can also construct the name of the enclosed CSV file. A temporary local filename is requested before we download the actual dataset. The dataset is unzipped and ingested into R. We then remove the downloaded file from the local storage.

```
url     <- ato_web_pkg$resources[[2]]$url
csvname <- ato_web_pkg$resources[[2]]$name %s+% ".csv"
temp    <- tempfile(fileext=".zip")

download.file(url, temp)
```

```
entry <- unz(temp, csvname) %>% read_csv()
unlink(temp)
```

As usual we obtain our first `dplyr::glimpse()` of the dataset.

```
glimpse(entry)
```

```
## Observations: 207,119
## Variables: 5
## $ Entry Page <chr> "http://www.ato.gov.au/", "http://www.a...
## $ Month      <chr> "Jul-13", "Jul-13", "Jul-13", "Jul-13",...
## $ Source     <chr> "External", "Internal", "External", "In...
## $ Views      <int> 9106745, 36423, 1991886, 1613, 1163224,...
## $ Visits     <int> 313151, 3643, 120417, 73, 167830, 41, 1...
```

Following our practise through the template we will load this dataset into our generic variable `ds`:

```
dsname <- "entry"
ds     <- get(dsname)
```

As we often note the variable names need to be simplified, or at least normalised using `rattle::normVarNames()`. We then record the list of available variables in `vars`.

```
names(ds) <- normVarNames(names(ds)) %T>% print()
```

```
## [1] "entry_page" "month"      "source"     "views"
## [5] "visits"
```

```
vars      <- names(ds)
```

The next task is to understand the character variables to determine whether they are better represented as factors. We review the distributions here.

```
ds %>%
  sapply(is.character) %>%
  which() %T>%
  {names(.) %>% print()} %>%
  select(ds, .) %>%
  sapply(function(x) unique(x) %>% length())
```

```
## [1] "entry_page" "month"       "source"
## entry_page      month     source
##      33262         11          3
```

Thus, `entry_page` would not be a candidate for conversion to a factor though `month` and `source` are. We consider each in turn.

For `month` we can construct a table of its distribution.

```
ds %>% select(month) %>% table()

## .
##   Apr-14   Aug-13   Dec-13 External   Feb-14   Jan-14   Ju...
##    23324    22849    19090        1    17222    17266    2...
##   Mar-14   Nov-13   Oct-13   Sep-13
##    18002    22478    22736    21084
```

Immediately we observe that the value `External` looks out of place. Checking the order of the columns in the dataset we note that this appears to be a value for the following column, `source`, rather than `month`. We appear to have a data issue.

To check this we first list the unique values of `month` and then view the observation with what appears to be an issue.

```
ds %>% select(month) %>% unique()

## # A tibble: 11 x 1
##       month
##       <chr>
##  1   Jul-13
##  2   Aug-13
##  3   Sep-13
....

ds %>% filter(month == "External") %>% print.data.frame()

##                                                    entry_page
## 1 http://www.ato.gov.au/content/00268103.htm ",March 2014"
##      month source views visits
## 1 External      4     3     NA
```

We can observe an odd value for `source` (4) and so on and

indeed notice that the date (`month`) is embedded within the `entry_page` character string.

Having identified a data quality issue we need to have a closer look to determine how widespread the issue is. Of course, initial indications are that there is a single bad observation in the dataset.

The first task is to identify where the errant observation occurs:

```
which(ds$source == "4")

## [1] 198499
```

With the observation number 198,499 in hand we can go back to the original source CSV file to check the source data itself. There are many ways to do this, including loading the file into a spreadsheet application or even just a text editor. A simple Linux command line approach will pipe the results of the Linux command `tail` into the command `head` as in the following command line (replacing the `<filename>` with the actual filename). Even for extremely large files, this will take almost no time at all as both commands are very efficient, and is likely quicker than loading the data into a spreadsheet.

```
$ tail -n+198500 <filename>.csv | head -n1
```

We discovery that row 198,500 (given that the first row of the CSV file is the header row) has the literal value:

```
"http://www.ato.gov.au/...268103.htm "",March 2014""",External,4,3,
```

Compare that to another row that is valid:

```
http://www.ato.gov.au/content/00171495.htm,Mar-14,External,7,2
```

Something would appear to have gone wrong in the extraction of the dataset at the source—the CSV file is the original from the web site so perhaps the extraction at the source was less than perfect or was damaged in transmission.

We must now decide what to do. We choose to simply `dplyr::filter()` out that observation. Alternatively we could have fixed the values for this observation within the dataset `ds` itself if it was felt that this observation needed to be included in the analysis or if the issue was found to be widespread.

```
dim(ds)

## [1] 207119      5

ds %<>% filter(month != "External")
dim(ds)

## [1] 207118      5
```

There is no further evidence on a review of the CSV file and the dataset of any other similar issues. We now convert this variable into a factor. We note that the order of the values of `month` as they appear in the original CSV file (and hence in the dataset) is nearly but not quite chronological so we must set the factor levels appropriately.

```
unique(ds$month)

##  [1] "Jul-13" "Aug-13" "Sep-13" "Oct-13" "Nov-13" "Dec-13"
##  [7] "Jan-14" "Feb-14" "Apr-14" "Mar-14"

months <- c("Jul-13", "Aug-13", "Sep-13", "Oct-13",
            "Nov-13", "Dec-13", "Jan-14", "Feb-14",
            "Mar-14", "Apr-14")
ds %<>% mutate(month=factor(month, levels=months))
```

We continue our observations noting that the variable `month` clearly records the month and year of an observation.

Our next character variable is the `source` which appears to record, presumably, whether the person browsing is external to the ATO or internal to the ATO. It has two distinct values (after removal of the errant observation above).

```
ds %>% select(source) %>% table()

## .
## External Internal
##   144400    62718
```

We decide that it is appropriate to convert this into a factor.

```
ds %<>% mutate(source=factor(source))
```

The remaining two variables are numeric. For an entry point the `views` appears to report the number of views for that month between internal and external views. Similarly for `visits`.

We summarise the dataset to get a feel for the shape of the data.

```
summary(ds)

##    entry_page              month                 source
##  Length:207118       Apr-14 :23324   External:144400
##  Class :character    Jul-13 :23067   Internal: 62718
##  Mode  :character    Aug-13 :22849
##                      Oct-13 :22736
##                      Nov-13 :22478
##                      Sep-13 :21084
##                      (Other):71580
##      views              visits
##  Min.   :      1   Min.   :      1.0
##  1st Qu.:      4   1st Qu.:      1.0
##  Median :     18   Median :      4.0
##  Mean   :   1019   Mean   :    169.2
##  3rd Qu.:     88   3rd Qu.:     20.0
##  Max.   :9106745   Max.   :1349083.0
##
```

The total number of views/visits to the website may be something of interest.

```
ds$views %>% sum %>% comcat()

## 211,110,271

ds$visits %>% sum %>% comcat()

## 35,050,249
```

We can then explore the views/visits per month. The following code produces the plot of Figure 6.3.

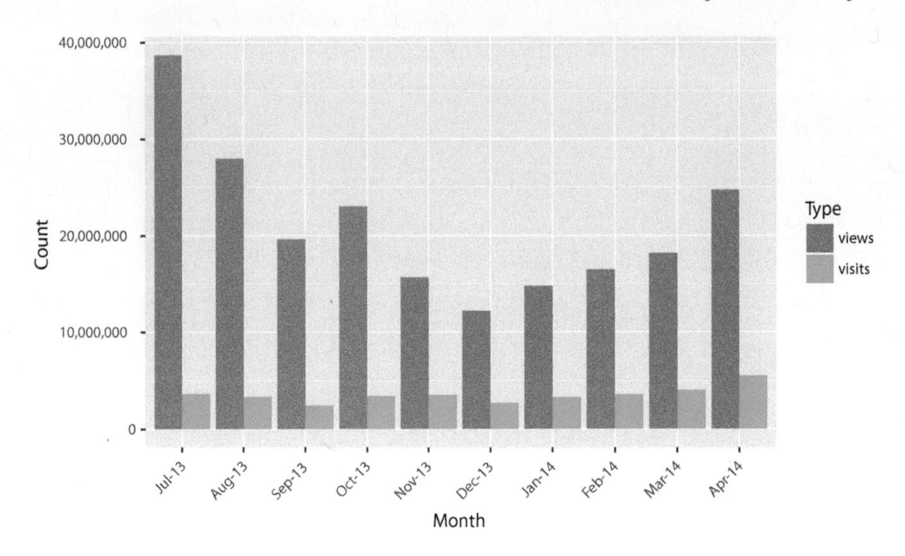

Figure 6.3: *Views and visits per month.*

```
ds %>%
  group_by(month) %>%
  summarise(views=sum(views), visits=sum(visits)) %>%
  gather(type, count, -month) %>%
  ggplot(aes(x=month, y=count, fill=type)) +
  geom_bar(stat="identity", position="dodge") +
  scale_y_continuous(labels=comma) +
  labs(fill="Type", x="Month", y="Count") +
  theme(axis.text.x=element_text(angle=45, hjust=1))
```

We may observe that there's a similar pattern over time for views and visits though to a different scale. The number of views (and also we would suggest views per visit) is dramatically increased for July. We need to analyse the relative change in views/visit over time to confirm the observation. Because of the different magnitudes between **views** and **visits** we might choose to use a log scale for the y-axis as in Figure 6.4. On the other hand, it may only serve to diminish the seasonal variation.

The July spike may well correspond to the Australian financial year ending in June and starting in July. We might also observe the holiday season around December when there must be less interest in taxation topics.

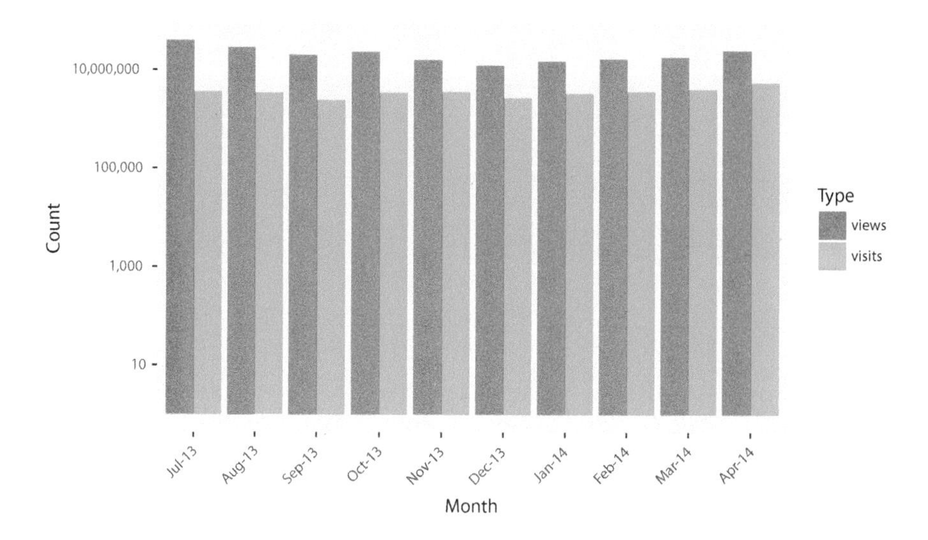

Figure 6.4: *Views and visits per month (log scale).*

Finally, a comparison of `External` and `Internal sources` may also be of interest. The following code segment generates the plot of Figure 6.5.

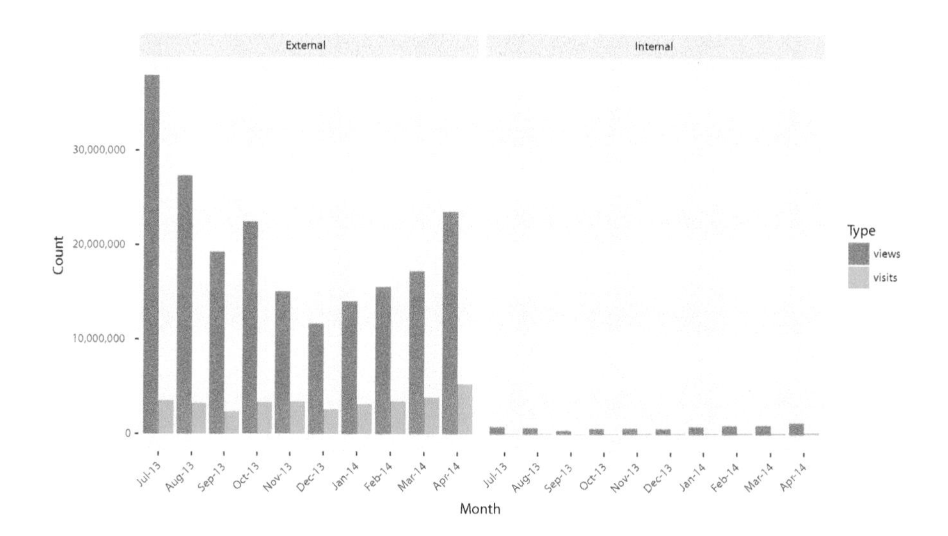

Figure 6.5: *Faceted plot of external and internal visits/views.*

```
ds %>%
  group_by(month, source) %>%
  summarise(views=sum(views), visits=sum(visits)) %>%
  gather(type, count, -c(month, source)) %>%
  ggplot(aes(x=month, y=count, fill=type)) +
  geom_bar(stat="identity", position="dodge") +
  scale_y_continuous(labels=comma) +
  labs(fill="Type", x="Month", y="Count") +
  theme(axis.text.x=element_text(angle=45, hjust=1)) +
  facet_wrap(~source)
```

The clear volume differences between external and internal visitors is not surprising with the ATO having only about 20,000 staff, compared to the population of Australia at over 23 million. Given this very stark differentiation between the external and internal populations, we might think to partition our analysis into the two cohorts. At a guess, we might expect the behaviours exhibited internally to be quite different to those exhibited through external accesses.

6.4 Exercises

Exercise 6.1 Exploring the Data

The full web log data suite contains multiple resources (datasets). Choose one that was not covered in this chapter and complete a study of the dataset, wrangling the data and extracting information and knowledge. Complete a report that tells the narrative of the dataset.

7

A Pattern for Predictive Modelling

In Chapter 3 we presented a process for ingesting, processing, reviewing, transforming and cleaning our data. There we introduced the concept of a template that captures through the use of generic variables a repeatable process. We can then use the template as a starting point for any new task. In this chapter we present a similar concept for building analytic or machine learning models.

The concept of building models was introduced in Chapter 1 where we highlighted the aim as capturing knowledge from data. Such knowledge represents an understanding in some form of the real world. Once the knowledge has been discovered from the data we can consider deploying the knowledge to reason about the world and indeed to build applications that intelligently interact with and further learn from its interaction with the world.

Many model builders, given a set of user specified parameters, use heuristics to search for a good model. It is typical in artificial intelligence for algorithms to perform a guided random search. The general approach is to search for a good model rather than to deterministically identify the best model. This is often necessary because the computational requirements to find the best model will generally be prohibitive and could take years of compute time.

Searching for the best model involves searching through an enormous search space of all of the possible expressions of a model in the particular knowledge representation language for the choice of model. We might compare model building to writing a sentence in a particular natural language (like English). The language has a particular specification or rules which are followed in building a sentence. Following these rules we can combine the words of the language together to form sentences. There is an infinite number of sentences we could form. As humans we search for the right sentence at the right time in the right context. We do this incredibly well.

Similarly for model building we have a language in which we express the model (the knowledge discovered). That language will typically allow for an infinite variety of sentences (models). Given a dataset (the context) our task is to search for the best or a good sentence that is most consistent with the data. Computers do this search automatically yet computationally because the search space is extremely large we cannot explore all possible sentences.

Heuristic algorithms reduce the computational requirements to something feasible. Such heuristics often involve some level of random decision making in deciding which paths to follow.

In this chapter we develop a pattern for building models using R. As with Chapter 3 the intention is to develop a pattern that can be used as the starting point for any future model building and evaluation of those models. By no means is the pattern the end point in itself. There will be plenty of variation to and extension of the patterns as we become more engaged with the model building for the particular characteristics of the dataset we are dealing with.

R offers a full suite of model builders. Each model builder is in practise a function call and the arguments to the function are generally similar across the different model builders. However, because they are written by different developers and at different times there are idiosyncratic differences between each. We need to sometimes be aware of these differences as we proceed.

At this time we have yet to see the development of a unified language for building models though the caret (Kuhn, 2017) package provides a well thought out and consistent interface to a large number of model builders. Our aim in this chapter is to provide a foundation for building and evaluating models in R and to begin with a simple template capturing the model building pattern. Once we have that foundation then the data scientist might like to become familiar with the advanced capabilities of caret (Kuhn, 2017).

Packages used in this chapter include ROCR (Sing *et al.*, 2015), dplyr (Wickham *et al.*, 2017a), ggplot2 (Wickham and Chang, 2016), magrittr (Bache and Wickham, 2014), randomForest

(Breiman *et al.*, 2015), rattle (Williams, 2017), rpart (Therneau *et al.*, 2017), scales (Wickham, 2016), stringi (Gagolewski *et al.*, 2017) and tibble (Müller and Wickham, 2017).

```
# Load required packages from the local library into R.

library(ROCR)         # Use prediction() for evaluation.
library(dplyr)        # select()
library(ggplot2)      # Visually evaluate performance.
library(magrittr)     # Data pipelines: %>% %T>% %<>%.
library(randomForest) #
library(rattle)       # Evaluate using riskchart(), comcat().
library(rpart)        # Model: decision tree.
library(scales)       # percent()
library(stringi)      # String operator: %s+%.
library(tibble)       # as_data_frame()
```

7.1 Loading the Dataset

In Chapter 3 we used `readr::read_csv()` to ingest our data into R. Some considerable effort then went into processing, reviewing, transforming and cleaning the data. We saved the dataset together with a collection of metadata as an R binary dataset file so that here we can `base::load()` the saved dataset and its associated metadata.

```
# Build the filename used to previously store the data.

fpath  <- "data"
dsname <- "weatherAUS"
dsdate <- "_20170702"
dsfile <- dsname %s+% dsdate %s+% ".RData"

fpath %>%
  file.path(dsfile) %T>%
  print() ->
dsrdata

## [1] "data/weatherAUS_20170702.RData"
```

```
# Load the R objects from file and list them.

load(dsrdata) %>% print()

##  [1] "ds"        "dsname"    "dspath"    "dsdate"
##  [5] "nobs"      "vars"      "target"    "risk"
##  [9] "id"        "ignore"    "omit"      "inputi"
## [13] "inputs"    "numi"      "numc"      "cati"
## [17] "catc"      "form"      "seed"      "train"
## [21] "validate"  "test"      "tr_target" "tr_risk"
....
```

The metadata can then be reviewed as a check of the veracity of the data. It is always important to regularly review our data. Errors can occur and can propagate through our modelling if we are not careful.

```
# Review the metadata.

dsname

## [1] "weatherAUS"

dspath

## [1] "http://rattle.togaware.com/weatherAUS.csv"

dsdate

## [1] "_20170702"

nobs %>% comcat()

## 134,776

vars

##  [1] "rain_tomorrow"    "min_temp"         "max_temp"
##  [4] "rainfall"         "evaporation"      "sunshine"
##  [7] "wind_gust_dir"    "wind_gust_speed" "wind_dir_9am"
## [10] "wind_dir_3pm"     "wind_speed_9am"   "wind_speed_3pm"
## [13] "humidity_9am"     "humidity_3pm"     "pressure_9am"
## [16] "cloud_9am"        "cloud_3pm"        "rain_today"
## [19] "season"           "cluster"
```

```
target

## [1] "rain_tomorrow"

risk

## [1] "risk_mm"

id

## [1] "date"     "location" "year"

ignore

## [1] "date"          "location"      "risk_mm"       "temp_3pm...
## [5] "pressure_3pm" "temp_9am"

omit

## NULL

train    %>% length() %>% comcat()

## 94,343

validate %>% length() %>% comcat()

## 20,216

test     %>% length() %>% comcat()

## 20,217
```

Notice specifically the training, validation, and testing "datasets". By building our model on the 70% training dataset we might expect it to well reflect that specific dataset. In fact, models can generally be quite accurate on the data on which they are trained. But this accuracy is a very optimistic (or biased) estimate of how the model generalises to other unseen datasets. Generally we are more interested to know how the model performs on unseen data and so how useful the model will be in a more general context. This is the role of the validation and testing datasets.

The validation and testing datasets are so-called ***hold-out*** datasets in that they have not been used at all for building the model. When we test the model on these datasets we would expect it to be less accurate. This is indeed what we will generally observe and we will see this in the following sections.

For our purposes we will illustrate the model building process with a specific type of model in mind—a binary classification model. Here the accuracy of a model relates to how accurate it can predict the target variable based on the input variables. Thus we often measure performance based on error rates.. The overall error rate measured on the training dataset will be shown to generally be less than the error rate calculated on the validation and testing datasets. These latter datasets provide an unbiased estimate of the true performance of the model.

7.2 Building a Decision Tree Model

There are many model building algorithms available to us in R. Indeed, essentially every machine learning algorithm is available in R in addition to R's traditional strength in statistical models.

A traditional and very popular machine learning algorithm is the decision tree induction algorithm. The target knowledge representation language is a structure referred to as a decision tree. The algorithm uses a divide and conquer approach to heuristically construct the decision tree. An information gain (entropy) or gini measure is typically used to guide the tree construction.

The popularity of this model is due in no small part to the choice of knowledge representation. As we will see shortly, the tree is readily understandable and presents the discovered knowledge quite clearly and in a form that non-technical collaborators can access.

A popular function to build decision tree models in R is `rpart::rpart()`. We will use this here to introduce the model building process.

To build a model we need to pass on to the model building

function the formula that describes the model to build. We have stored the formula within the generic variable **form**. We also pass on to the algorithm the historic dataset from which we will build the model. The training dataset as a subset of the full dataset **ds** is identified along with just the **vars** (variables or columns) that we will model. Using generic variables allows us to change the formula, the dataset, the observations and the variables used in building the model yet retain the same function call.

```
# Train a decision tree model.

m_rp  <- rpart(form, ds[train, vars])
```

The result of the model build is stored into the variable **m_rp**. If we are experimenting with different model build parameters we might have several models that we wish to save, and might save them as **m_rp_01**, **m_rp_02**, and so on.

In line with our template approach we introduce generic variables to store the actual model and meta data about the model.

```
# Initialise generic variables.

model <- m_rp
mtype <- "rpart"
mdesc <- "decision tree"
```

The generic variable **model** will be used to refer to the model in a general way. We also introduce the generic variable **m_type** to record the type of the model we have built and **m_desc** as a human readable description of the model type. We will see these used later.

We can view the model by referencing the generic variable **model** on the command line. R will base::**print()** the model to the console output.

```
# Basic model structure.

model

## n= 94343
```

```
##
## node), split, n, loss, yval, (yprob)
##       * denotes terminal node
##
##  1) root 94343 21244 no (0.7748217 0.2251783)
##    2) humidity_3pm< 71.5 78710 11493 no (0.8539830 0.14601...
##    3) humidity_3pm>=71.5 15633  5882 yes (0.3762554 0.6237...
##      6) humidity_3pm< 83.5 8977  4338 no (0.5167651 0.4832...
##       12) wind_gust_speed< 47 6290  2587 no (0.5887122 0.4...
##         24) rainfall< 2.3 4295  1479 no (0.6556461 0.34435...
##         25) rainfall>=2.3 1995   887 yes (0.4446115 0.5553...
##       13) wind_gust_speed>=47 2687   936 yes (0.3483439 0....
##      7) humidity_3pm>=83.5 6656  1243 yes (0.1867488 0.813...
```

This textual version of the model provides the basic structure of the tree. Different model builders will base::**print()** different information. We get a glimpse of the knowledge discovered here, represented as a tree.

We can also review the base::**summary()** of the model build process.

```
# Basic model build summary.

summary(model)

## Call:
## rpart(formula=form, data=ds[train, vars])
##   n= 94343
##
##           CP nsplit rel error    xerror        xstd
## 1 0.18212201      0 1.0000000 1.0000000 0.006039246
## 2 0.02626624      1 0.8178780 0.8178780 0.005604365
## 3 0.01040294      3 0.7653455 0.7655809 0.005461220
## 4 0.01000000      4 0.7549426 0.7641687 0.005457229
##
## Variable importance
##    humidity_3pm        max_temp wind_gust_speed
##              87               3               3
##        rainfall      rain_today  wind_speed_3pm
##               2               1               1
##    humidity_9am  wind_speed_9am         cluster
##               1               1               1
##
```

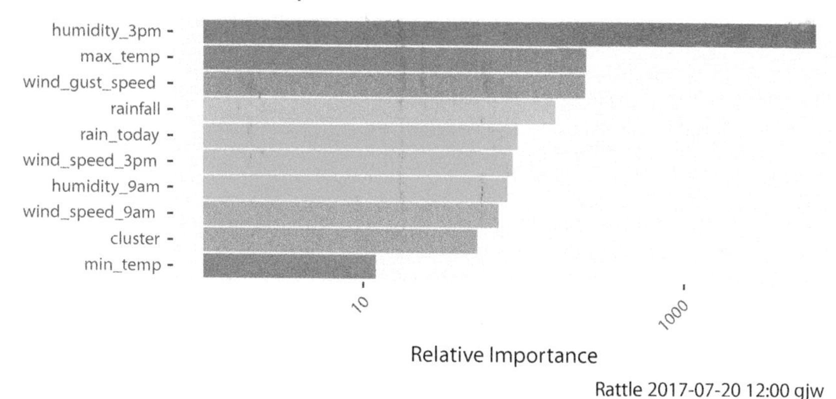

Figure 7.1: *Decision tree variable importance.*

```
## Node number 1: 94343 observations,    complexity param=0.1...
##   predicted class=no   expected loss=0.2251783  P(node) =1
....
```

A primary goal of data science is to gain insight into our data. As with artificial intelligence and machine learning we often want to review the discovered knowledge. The above textual representation of the discovered knowledge can be augmented with a visualisation of the same knowledge. A number of tools are available to support the visualisation of our models and to communicate understanding of the knowledge discovered through the model building process.

Often we are interested in knowing which variables were important in building the model to predict the outcome. We can use `rattle::ggVarImp()` to plot the variable importance as in Figure 7.1. Note the use of `log=TRUE` in the code below to force the axis to be a log scale so as to show the details for variables of less importance.

```
# Review variable importance.

ggVarImp(model, log=TRUE)
```

A visual representation of the decision tree model itself is also

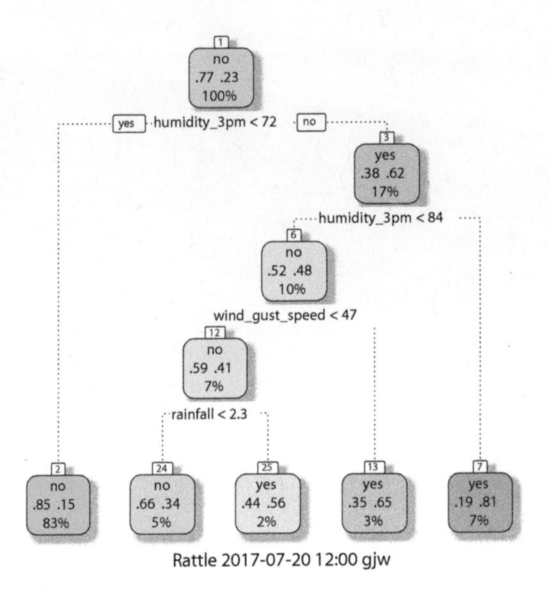

Figure 7.2: *Decision tree visualisation.*

quite useful in exposing the structure or knowledge discovered from the dataset. We use `rattle::fancyRpartPlot()` for a visually interesting presentation of the tree in Figure 7.2

```
# Visualise the discovered knowledge.

fancyRpartPlot(model)
```

We can quickly grasp the knowledge from the visual representation. Briefly, the path travelling through the left-hand side of the tree essentially captures that if the humidity at 3pm today was less than 72 then we expect there to be no rain tomorrow with an 85% probability. Taking the rightmost path through the tree, the rules predict that it will rain with a 79% probability if the humidity at 3pm is greater than 82.

This then is our first predictive analytic model built using a machine learning algorithm. In particular, it is a model that exposes the discovered knowledge from the provided dataset.

7.3 Model Performance

Having built a predictive analytic model we will want to understand how well the model performs. We will now apply the model to the observations in the training dataset to evaluate its performance. A number of measures will be introduced and many more exist. The choice depends on the data and the task at hand.

Predict

The generic function used in R to obtain predictions from a model is `stats::predict()`. We can obtain the predictions on the training dataset to evaluate the accuracy of the model.

For different types of models `stats::predict()` will generally behave in a similar and consistent way. There are however variations that we need to be aware of as we use new machine learning algorithms. For a `decision tree` model to produce class predictions for a dataset of observations we specify the `type=` of prediction as `"class"`:

```
# Predict on the training dataset.

model %>%
  predict(newdata=ds[train, vars], type="class") %>%
  set_names(NULL) %T>%
  {head(., 20) %>% print()} ->
tr_class
```

```
##  [1] no  no  no  no  no  no  no  no  no  no  no  no  no  no
## [15] yes no  no  no  no  no
## Levels: no yes
```

We can then compare this to the actual class for these observations as is recorded in the original training dataset. The actual classes have already been stored as the variable `tr_target`:

```
head(tr_target, 20)
```

```
##  [1] no  no  no  no  no  no  no  no  no  no  no  no  no  no
## [15] no  no  no  no  yes no
## Levels: no yes
```

Observe that the model correctly predicts 18 of the first 20 training dataset observations and 78,305 of the 94,343 training observations. We perform these calculations in the following code block. Notice the use of base::sum() applied to a vector of Boolean values (TRUE/FALSE). R automatically treats TRUE as 1 and FALSE as 0 so that we base::sum() the number of TRUE values in the vector.

```
head(tr_class) == head(tr_target)

## [1] TRUE TRUE TRUE TRUE TRUE TRUE

sum(head(tr_class) == head(tr_target))

## [1] 6

sum(tr_class == tr_target)

## [1] 78305
```

For a decision tree model we can also obtain the probability of it raining tomorrow using stats::predict() with type="prob".

```
model %>%
  predict(newdata=ds[train, vars], type="prob") %>%
  .[,2] %>%
  set_names(NULL) %>%
  round(2) %T>%
  {head(., 20) %>% print()} ->
tr_prob

##  [1] 0.15 0.15 0.15 0.15 0.15 0.15 0.15 0.34 0.15 0.15 0.15
## [12] 0.15 0.15 0.15 0.65 0.15 0.15 0.15 0.15 0.15
```

Accuracy and Error Rate

Given the predicted classes and probabilities we are now in a position to evaluate the performance of the model. The simplest measure is to calculate how accurate the model is over all of the predictions. We saw this calculation above but now convert it to a percentage accuracy.

```
sum(tr_class == tr_target) %>%
  divide_by(length(tr_target)) %T>%
  {
    percent(.) %>%
    sprintf("Overall accuracy = %s\n", .) %>%
    cat()
  } ->
tr_acc

## Overall accuracy=83%
```

Similarly we can calculate the model error rate which is sometimes used in preference to the accuracy.

```
sum(tr_class != tr_target) %>%
  divide_by(length(tr_target)) %T>%
  {
    percent(.) %>%
    sprintf("Overall error = %s\n", .) %>%
    cat()
  } ->
tr_err

## Overall error=17%
```

The model has an overall accuracy of 83% and error rate of 17%. That is a relatively high accuracy for a typical model build.

Confusion Matrix

The accuracy and error are rather blunt measures of performance. They are a good starting point to get a sense of how good the model is but more is required. A confusion matrix allows us to review how well the model performs against actual classes.

We begin by counting the number of times the model agrees with the actual data across the different classes. The comparison is made on the training dataset comparing the target (`tr_target`) to the predicted class (`tr_class`). Such a table is referred to as a confusion or error matrix.

```
# Basic comparison of prediction and actual as a confusion matrix.

table(tr_target, tr_class, dnn=c("Actual", "Predicted"))

##        Predicted
## Actual     no   yes
##     no  70033  3066
##     yes 12972  8272
```

It is often most useful to convert these counts to rates or percentages.

```
# Calculate percentages for confusion matrix.

table(tr_target, tr_class, dnn=c("Actual", "Predicted")) %>%
  divide_by(length(tr_target)) %>%
  multiply_by(100) %>%
  round(1)

##        Predicted
## Actual    no  yes
##     no  74.2  3.2
##     yes 13.7  8.8
```

We will find ourselves repeating the use of this formulation often to produce a summary of the performance of our model. As such we would be inclined to define our own function to do this so we do not need to repeat ourselves. In fact though this formulation is implemented as `rattle::errorMatrix()` which reports the percentages as well as the class error rates. It includes options to report the raw counts or, by default, the percentages. We store the resulting matrix of percentages as the variable `tr_matrix` for later usage.

```
# Count of predictions across all observations.

errorMatrix(tr_target, tr_class, count=TRUE)

##        Predicted
## Actual     no  yes Error
##     no  70033 3066   4.2
##     yes 12972 8272  61.1
```

```
# Comparison as percentages of all observations.

errorMatrix(tr_target, tr_class) %T>%
  print() ->
tr_matrix

##        Predicted
## Actual   no yes Error
##     no  74.2 3.2   4.2
##     yes 13.7 8.8  61.1
```

Common terminology refers to 74.2 as the true negatives, 3.2 as the false positives, 13.7 as the false negatives and 8.8 as the true positives.

Using `tr_matrix` we can calculate the error rate and an average of the class error rates.

```
tr_matrix %>%
  diag() %>%
  sum(na.rm=TRUE) %>%
  subtract(100, .) %>%
  sprintf("Overall error percentage = %s%%\n", .) %>%
  cat()

## Overall error percentage=17%

tr_matrix[,"Error"] %>%
  mean(na.rm=TRUE) %>%
  sprintf("Averaged class error percentage = %s%%\n", .) %>%
  cat()

## Averaged class error percentage=32.65%
```

A confusion matrix is particularly useful when the consequences of a wrong decision are different for the different decisions. For example, if it is incorrectly predicted that it will not rain tomorrow and I decide not to carry an umbrella with me then as a consequence I will get wet. We might experience this as a more severe consequence than the situation where it is incorrectly predicted that it will rain and so I unnecessarily carry an umbrella with me all day. The significance of different error types in medicine (predicting cancer) can be even more dramatic.

Three further measures of performance are introduced based on the confusion matrix: the recall, precision, and F-score. The recall is the proportion of true positives that are identified by the model. For the weather dataset that corresponds to the proportion of days that it rains and the model predicts it will rain. The precision is the proportion of true positives that are amongst the positives predicted by the model. For the weather dataset this is the proportion of the number of days for which it is predicted to rain and it does actually rain. Larger values for both are better. The F-score is calculated as the harmonic mean of these two measures.

```
# Recall.

tr_rec <- (tr_matrix[2,2]/(tr_matrix[2,2]+tr_matrix[2,1])) %T>%
  {percent(.) %>% sprintf("Recall = %s\n", .) %>% cat()}

## Recall=39.1%

# Precision.

tr_pre <- (tr_matrix[2,2]/(tr_matrix[2,2]+tr_matrix[1,2])) %T>%
  {percent(.) %>% sprintf("Precision = %s\n", .) %>% cat()}

## Precision=73.3%

# F-Score.

tr_fsc <- ((2 * tr_pre * tr_rec)/(tr_rec + tr_pre))  %T>%
  {sprintf("F-Score = %.3f\n", .) %>% cat()}

## F-Score=0.510
```

7.3.1 ROC Curve

Another common measure of the performance of a model is the ROC curve and in particular the area under the ROC curve. This area can be calculated using `ROCR::prediction()` and `ROCR::performance()`. These functions use the probability of a prediction rather than the prediction of a class.

In the following code block we obtain the predicted probabilities from the model, predicting over the training dataset. The result from stats::predict() for an rpart model is a matrix with columns corresponding to the possible class values recording the probability of each class for each observation. The second column is the one of interest (the probability that it will rain tomorrow, i.e., **rain_tomorrow==yes**). These probabilities are passed on to ROCR::prediction() to compare them with the actual target values. The result is then passed on to ROCR::performance() from which we obtain the base::attr()ibute **y.values** and then extract the first value as the area under the curve.

```
tr_prob %>%
  prediction(tr_target) %>%
  performance("auc") %>%
  attr("y.values") %>%
  .[[1]] %T>%
  {
    percent(.) %>%
    sprintf("Percentage area under the ROC curve = %s\n", .) %>%
    cat()
  } ->
tr_auc
```

```
## Percentage area under the ROC curve=69.8%
```

The area under the curve (AUC) is 69.8% of the total area. If it were 100% then the model would be perfectly accurate and usually an indication that the model has over-fit the data and so will not perform well on new data.

The area under the curve can be visualized if we draw the actual ROC curve. To do so we first need to calculate a true positive rate and a false positive rate using ROCR::performance(). The two values are abbreviated as **tpr** and **fpr**, respectively.

```
tr_prob %>%
  prediction(tr_target) %>%
  performance("tpr", "fpr") ->
tr_rates
```

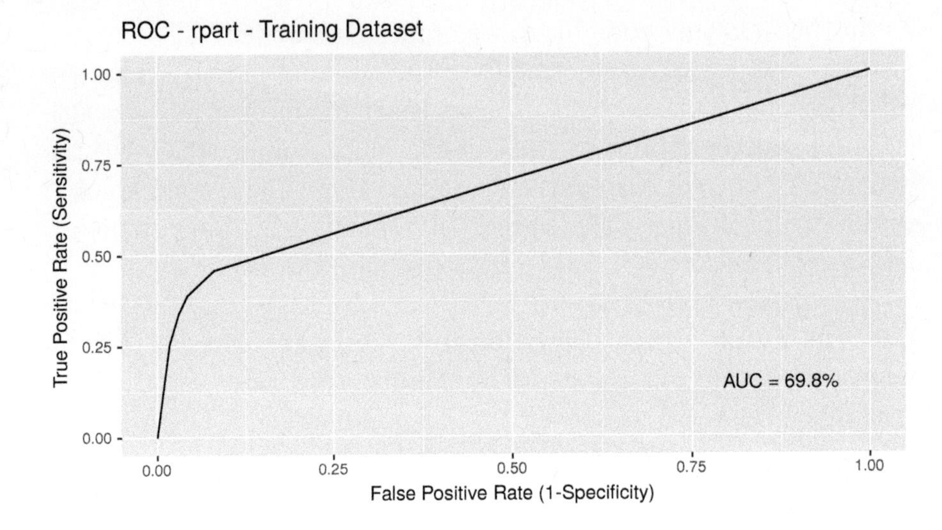

Figure 7.3: *ROC curve for decision tree over training dataset. The area under the curve is a measure of the performance of the model. A perfect model would have 100% of the area under the curve.*

The false positive rate is calculated to be used as the x values and the true positive rate as the y values. Extracting these into a data frame will allow us to pass the data on to `ggplot2::ggplot()`. A line is shown for the ROC curve and we add some explanatory labels and text to report the calculated area under the curve. The resulting plot is shown in Figure 7.3.

```
data_frame(tpr=attr(tr_rates, "y.values")[[1]],
           fpr=attr(tr_rates, "x.values")[[1]]) %>%
  ggplot(aes(fpr, tpr)) +
  geom_line() +
  annotate("text", x=0.875, y=0.125, vjust=0,
           label=paste("AUC =", percent(tr_auc))) +
  labs(title = "ROC - " %s+% mtype %s+% " - Training Dataset",
       x     = "False Positive Rate (1-Specificity)",
       y     = "True Positive Rate (Sensitivity)")
```

The more area under the curve the better the model is performing.

7.3.2 Risk Chart

A risk chart (Williams, 2011), also known as an accumulative performance plot, is implemented in rattle as an alternative to an ROC curve. It can be useful for explaining the performance of a model. For many financial audit type environments, for example, business is expressed in terms of audit cases and case loads. That is, a case load or population of customers are to be audited for financial compliance. Some percentage will be identified as non-compliant (the true positives). The remainder will have been audited unnecessarily (the false positives). The ROC curve is useful in presenting the trade-off between the true positives (the y-axis) and the false positives (the x-axis). A risk chart plots the true positive rate against the caseload (the total number of positives which is the combined true and false positives).

Figure 7.4 shows a `rattle::riskchart()` based on the training dataset. The x-axis can be considered as a case load whilst the y-axis is the performance over that case load. In approximate terms we can read, for example, that with a case load of 40% we have a return of 60% of the successful cases. Of course the data here relates to our weather data but the interpretation carries through to audit data.

```
riskchart(tr_prob, tr_target, tr_risk) +
  labs(title="Risk Chart - " %s+%
         mtype %s+%
         " - Training Dataset") +
  theme(plot.title=element_text(size=14))
```

Just as with a ROC curve the more area under the curve the better the model performance. The solid grey line shows where the the perfect performance would be.

7.4 Evaluating Model Generality

Having an understanding of how well the model performs on the data on which it was built, we might ask how will the model per-

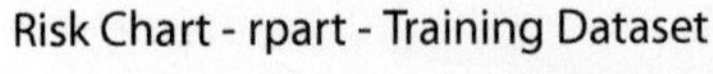

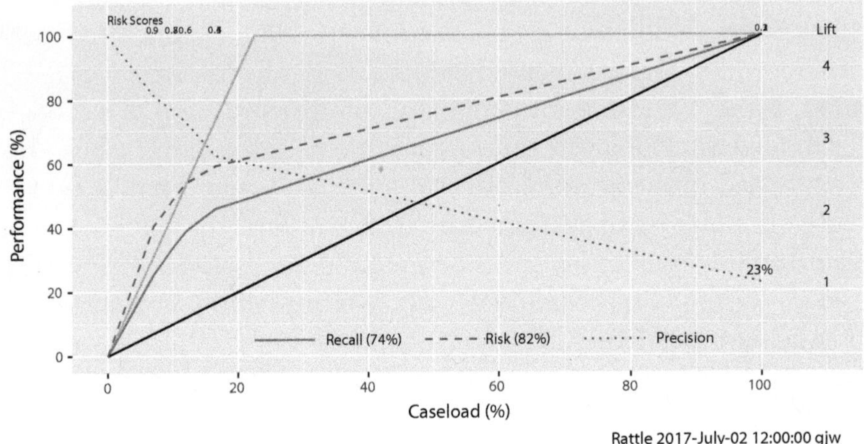

Figure 7.4: *A risk chart for the training dataset. The risk chart plots the population against performance, captured as the true positive rate. The area under the risk chart is another single measure for comparing performance.*

form on new data. This is the purpose of the validation dataset. We evaluate a model by making predictions on observations that were not used in building the model. These observations will need to have a known outcome so that we can compare the model prediction against the known outcome. Generally the evaluation of a model over the training dataset will result in a biased estimate of the actual model performance—after all, that is the data that was used to build the model and so the model builder should do well on that dataset.

We now introduce the concept of making predictions over new, unseen, or hold-out observations using the model we have built. We will use these predictions over the validation dataset to provide an unbiased estimate of the quality of the predictions made by the model. This may lead us to fine-tune some of the parameters we have available for building a decision tree, such as the maximum depth, the complexity, and so on.

We will replicate the evaluation performed on the training dataset but now on the validation dataset.

Predict

```
# Predict on the validation dataset.

model %>%
  predict(newdata=ds[validate, vars], type="class") %>%
  set_names(NULL) %T>%
  {head(., 20) %>% print()} ->
va_class

##  [1] yes no  no  no  no  no  no  no  no  yes no  no  no  no
## [15] no  no  no  no  no  no
## Levels: no yes
```

Again we compare the predictions to the actual class for these observations.

```
head(va_target, 20)

##  [1] yes no  no  no  no  no  no  no  no  no  yes no  no  no
## [15] no  yes no  no  no  yes
## Levels: no yes
```

The model correctly predicts 16 of the first 20 validation dataset observations and 16,780 of the 20,216 validation observations.

We will also record the probability of it raining tomorrow.

```
model %>%
  predict(newdata=ds[validate, vars], type="prob") %>%
  .[,2] %>%
  set_names(NULL) %>%
  round(2) %T>%
  {head(., 20) %>% print()} ->
va_prob

##  [1] 0.65 0.15 0.15 0.15 0.15 0.15 0.15 0.15 0.34 0.81 0.34
## [12] 0.15 0.15 0.15 0.15 0.15 0.15 0.15 0.15 0.15
```

Accuracy and Error Rate

The accuracy and error over the validation dataset can be calculated as previously for the training dataset.

```
sum(va_class == va_target) %>%
  divide_by(length(va_target)) %T>%
  {
    percent(.) %>%
      sprintf("Overall accuracy = %s\n", .) %>%
      cat()
  } ->
va_acc

## Overall accuracy=83%
```

```
sum(va_class != va_target) %>%
  divide_by(length(va_target)) %T>%
  {
    percent(.) %>%
      sprintf("Overall error = %s\n", .) %>%
      cat()
  } ->
va_err

## Overall error=17%
```

The model has an overall accuracy of 83% and error rate of 17%. For our simple model here these are the same as with the trianing dataset.

Confusion Matrix

The counts of the agreements between the model and the actual values for the validation dataset is the first step to computing the confusion matrix based on the validation dataset comparing the target (`va_target`) to the predicted class (`va_class`).

```
# Basic comparison of prediction/actual as a confusion matrix.

errorMatrix(va_target, va_class, count=TRUE)
```

```
##        Predicted
## Actual    no  yes Error
##     no  15078  662   4.2
##     yes  2774 1702  62.0
```

The actual percentages can also be obtained.

```
# Comparison as percentages of all observations.

errorMatrix(va_target, va_class) %T>%
  print() ->
va_matrix
```

```
##        Predicted
## Actual   no yes Error
##     no  74.6 3.3   4.2
##     yes 13.7 8.4  62.0
```

The error matrix saved to `va_matrix` is used to calculate the error rate and an average of the class error rates.

```
va_matrix %>%
  diag() %>%
  sum(na.rm=TRUE) %>%
  subtract(100, .) %>%
  sprintf("Overall error percentage = %s%%\n", .) %>%
  cat()
```

```
## Overall error percentage=17%
```

```
va_matrix[,"Error"] %>%
  mean(na.rm=TRUE) %>%
  sprintf("Averaged class error percentage = %s%%\n", .) %>%
  cat()
```

```
## Averaged class error percentage=33.1%
```

Similarly for recall, precision, and the F-score.

```
va_rec <- (va_matrix[2,2]/(va_matrix[2,2]+va_matrix[2,1])) %T>%
  {percent(.) %>% sprintf("Recall = %s\n", .) %>% cat()}
```

```
## Recall=38%
```

```
va_pre <- (va_matrix[2,2]/(va_matrix[2,2]+va_matrix[1,2])) %T>%
  {percent(.) %>% sprintf("Precision = %s\n", .) %>% cat()}

## Precision=71.8%

va_fsc <- ((2 * va_pre * va_rec)/(va_rec + va_pre))  %T>%
  {sprintf("F-Score = %.3f\n", .) %>% cat()}

## F-Score=0.497
```

Compare the performance measured by the validation dataset to that on the training dataset. The very minor increased accuracy is unlikely to be statistically significant but in general it is not unusual to have a more optimistic performance on the training dataset.

```
tr_matrix

##         Predicted
## Actual    no yes Error
##      no  74.2 3.2   4.2
##      yes 13.7 8.8  61.1

va_matrix

##         Predicted
## Actual    no yes Error
##      no  74.6 3.3   4.2
##      yes 13.7 8.4  62.0
```

7.4.1 ROC Curve

The ROC curve and associated area under the ROC curve are again calculated over the validation dataset. We see the ROC curve in Figure 7.5.

```
va_prob %>%
  prediction(va_target) %>%
  performance("auc") %>%
  attr("y.values") %>%
```

```
  .[[1]] %T>%
  {
    percent(.) %>%
    sprintf("Percentage area under the ROC curve = %s\n", .) %>%
    cat()
  } ->
va_auc

## Percentage area under the ROC curve=69.5%
```

```
va_prob %>%
  prediction(va_target) %>%
  performance("tpr", "fpr") ->
va_rates
```

```
data_frame(tpr=attr(va_rates, "y.values")[[1]],
           fpr=attr(va_rates, "x.values")[[1]]) %>%
  ggplot(aes(fpr, tpr)) +
  geom_line() +
  annotate("text", x=0.875, y=0.125, vjust=0,
           label=paste("AUC =", percent(va_auc))) +
  labs(title="ROC - " %s+% mtype %s+% " - Validation Dataset",
       x="False Positive Rate (1-Specificity)",
       y="True Positive Rate (Sensitivity)")
```

7.4.2 Risk Chart

Figure 7.6 shows the `rattle::riskchart()` based on the validation dataset.

```
riskchart(va_prob, va_target, va_risk) +
  labs(title="Risk Chart - " %s+%
         mtype %s+%
         " - Validation Dataset") +
  theme(plot.title=element_text(size=14))
```

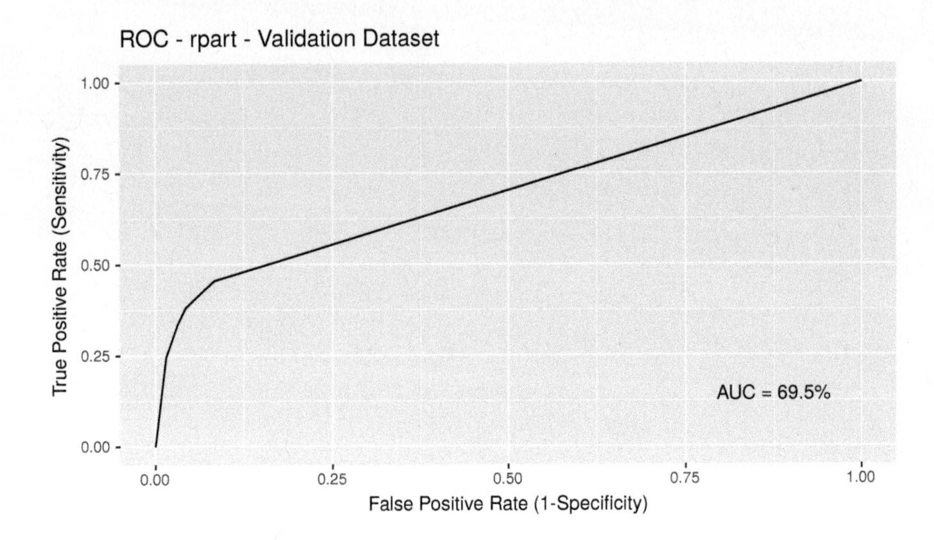

Figure 7.5: *ROC curve for decision tree over validation dataset.*

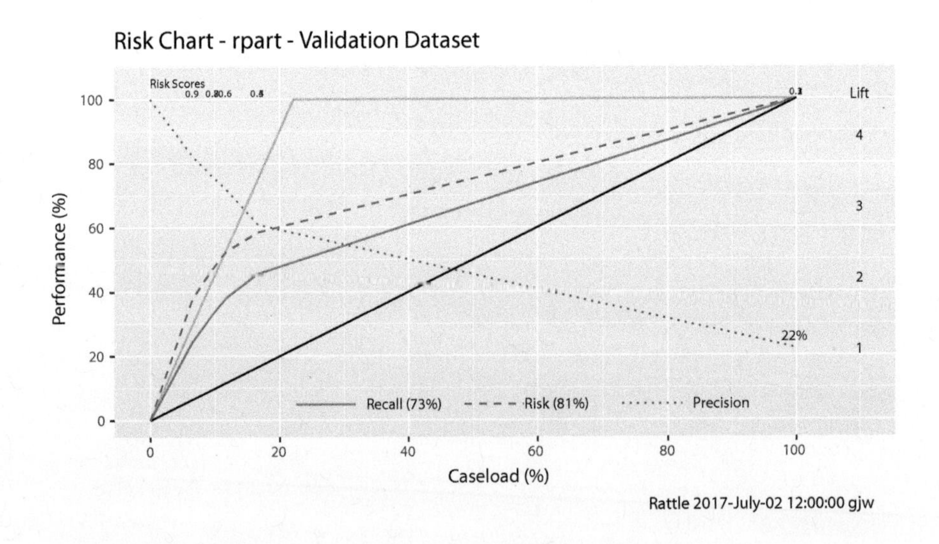

Figure 7.6: *A Risk Chart for the validation dataset. The risk chart plots the population against performance, captured as the true positive rate. The area under the risk chart is another single measure for comparing performance.*

7.5 Model Tuning

Once we have built our first model the journey begins to tune the parameters of the model building algorithm to explore for the model that best fits the data. There are tools available that will search through a series of parameter values, building a model for each setting, and identifying the combination of parameters that leads to the best model. We will illustrate the process here as a manual tuning step.

As with most machine learning algorithms there are many options available to affect the model that is built by the function. For `rpart::rpart()` we have options available to change the maximum number of nodes down a single path of the tree (`maxdepth=` wiht 30 as the default), the minimum number of observations in a node before we would consider splitting the node further (`minsplit=`20), the minimum number of observations within any terminal (leaf) node (`minbucket=`7), and many more. A particularly important argument is complexity parameter (`cp=`0.01). Smaller values will tend to build larger trees by allowing splits to a node within the tree for smaller gains to the accuracy of the model. See the manual page for `rpart::rpart.control()` for further details.

Here we use for illustrative purposes a set of parameters that build a considerably more complex decision tree model.

```
# Build a more complex model

control <- rpart.control(cp=0.0001)
m_rp01  <- rpart(form, ds[train, vars], control=control)

# Copy model into generic variables.

model <- m_rp01
mtype <- "rpart"
mdesc <- "decision tree (cp 1e-4)"

model

## n= 94343
```

```
##
## node), split, n, loss, yval, (yprob)
##       * denotes terminal node
##
##       1) root 94343 21244 no (0.77482166 0.22517834)
##         2) humidity_3pm< 71.5 78710 11493 no (0.85398298 0...
##           4) sunshine>=7.75 52590  4389 no (0.91654307 0.0...
##             8) wind_gust_speed< 53 46015  3132 no (0.93193...
##              16) humidity_3pm< 53.5 37071  1918 no (0.9482...
##              17) humidity_3pm>=53.5 8944  1214 no (0.86426...
##                34) pressure_9am>=1014.35 6607   700 no (0....
##                  68) cloud_3pm< 2.5 2458   187 no (0.92392...
##                   136) wind_gust_speed< 42 2084   131 no (...
##                   137) wind_gust_speed>=42 374    56 no (0...
##                     274) cluster=area2,area3,area5 354    ...
##                     275) cluster=area1,area4 20      7 yes ...
##                       550) wind_speed_3pm>=23 7      2 no (...
##                       551) wind_speed_3pm< 23 13      2 yes...
##                  69) cloud_3pm>=2.5 4149   513 no (0.87635...
....
```

This is an excessively complex decision tree. Whereas our first model had 9 nodes this new model has 1,519 nodes. This is considerably more complex and we now ask whether the additional complexity leads to performance gains.

The accuracy of the model will be assessed on the training and validation datasets.

```
# Use the model to predict.

model %>%
  predict(newdata=ds[train, vars], type="class") %>%
  set_names(NULL) %T>%
  {head(., 20) %>% print()} ->
tr_class

##  [1] no  no  no  no  no  no  no  no  no  no  no  no  no  no
## [15] no  no  no  no  yes no
## Levels: no yes

model %>%
  predict(newdata=ds[validate, vars], type="class") %>%
  set_names(NULL) %T>%
```

```
  {head(., 20) %>% print()} ->
va_class

## [1] yes no  no  no  no  no  no  yes yes yes no  no  no  no
## [15] no  yes no  no  no  no
## Levels: no yes


# Compare model accuracy.

sum(tr_class == tr_target) %>%
  divide_by(length(tr_target)) %>%
  percent(.)

## [1] "87.8%"

sum(va_class == va_target) %>%
  divide_by(length(va_target)) %>%
  percent(.)

## [1] "83.6%"
```

Notice that the accuracy of model on the training dataset is higher than on the data the model has not previously been exposed to. The reported performance of the model on the validation dataset is likely to be more in line with how well the model will perform on any new data. That is, the measure on the validation dataset is a less biased estimate of the true performance of the model. We can see a hint of the issue of over-fitting here.

We observe a similar pattern for the error matrix.

```
# Compare model performance across all three datasets.

errorMatrix(tr_target, tr_class)

##       Predicted
## Actual   no  yes Error
##    no  74.6  2.9   3.8
##    yes  9.3 13.2  41.3

errorMatrix(va_target, va_class)
```

```
##        Predicted
## Actual   no  yes Error
##    no  72.8  5.1   6.5
##    yes 11.3 10.8  51.3
```

As noted previously for our scenario false negatives (when we fail to predict that it will rain) have more impact than the false positives (predicting that it will rain yet it does not). Reviewing the error matrix and in particular the class error rates (the final column) we clearly have a problem with too many false negatives.

A useful parameter to address this situation is the `loss=` parameter of the `parms=` argument. This is used to indicate that not all outcomes are equal. In the following example we specify that the false negatives are a greater loss than the false positives.

```
# Bias model build towards reducing false negatives.

loss   <- matrix(c(0,1,10,0), byrow=TRUE, nrow=2)
m_rp02 <- rpart(form, ds[train, vars], parms=list(loss=loss))

# Copy model into generic variables.

model <- m_rp02
mtype <- "rpart"
mdesc <- "decision tree (loss fn = 10*fp)"

model

## n= 94343
##
## node), split, n, loss, yval, (yprob)
##       * denotes terminal node
##
##  1) root 94343 73099 yes (0.77482166 0.22517834)
##    2) humidity_3pm< 58.5 59403 53371 yes (0.89845631 0.101...
##      4) sunshine>=9.35 35324 23060 no (0.93471860 0.065281...
##        8) pressure_9am>=1014.45 24684 11370 no (0.95393777...
##        9) pressure_9am< 1014.45 10640  9471 yes (0.8901315...
##         18) cloud_3pm< 2.5 5482  3730 no (0.93195914 0.068...
##         19) cloud_3pm>=2.5 5158  4362 yes (0.84567662 0.15...
##      5) sunshine< 9.35 24079 20353 yes (0.84525935 0.15474...
##       10) pressure_9am>=1021.85 9096  6050 no (0.93348725 ...
```

```
##        11) pressure_9am< 1021.85 14983 11862 yes (0.7916972...
##     3) humidity_3pm>=58.5 34940 19728 yes (0.56462507 0.435...
```

```
model %>%
  predict(newdata=ds[validate, vars], type="class") %>%
  set_names(NULL) %T>%
  {head(., 20) %>% print()} ->
va_class
```

```
##  [1] yes no  yes yes no  yes no  yes yes yes yes yes yes yes
## [15] yes yes yes yes no  yes
## Levels: no yes
```

```
sum(va_class == va_target) %>%
  divide_by(length(va_target)) %>%
  percent(.)
```

```
## [1] "59.3%"
```

```
errorMatrix(va_target, va_class)
```

```
##        Predicted
## Actual    no  yes Error
##     no  39.3 38.6  49.6
##     yes  2.1 20.0   9.6
```

We have dramatically reduced the false negative rate and improved the class error for the positive class (it rains tomorrow) significantly. The overall error rate has increased which is okay since we have purposefully identified that the consequences of a false positive (we carry an umbrella around for the day) are less than for a false negative (we get wet).

This might now be the model we decide to proceed with into production perhaps as an app that will collect today's weather data and then provides advice each morning as to whether we should pack an umbrella for the day.

We would finally like to understand how accurate the model is expected to be noting that now we have actually used the validation dataset in the process of building the model. We now bring in the testing dataset which has so far not been used at all for any model evaluation. Using the testing dataset we proceed through our evaluation script. We expect these performance measures to be an unbiased estimate of the performance of the model on new data.

```
model %>%
  predict(newdata=ds[test, vars], type="class") %>%
  set_names(NULL) %T>%
  {head(., 20) %>% print()} ->
te_class

##  [1] yes no  no  no  no  no  yes no  no  no  no  yes yes no
## [15] no  yes no  no  no  no
## Levels: no yes

model %>%
  predict(newdata=ds[test, vars], type="prob") %>%
  .[,2] %>%
  set_names(NULL) %>%
  round(2) %T>%
  {head(., 20) %>% print()} ->
te_prob

##  [1] 0.15 0.07 0.07 0.07 0.07 0.05 0.44 0.05 0.05 0.05 0.05
## [12] 0.44 0.15 0.05 0.05 0.21 0.05 0.05 0.05 0.05

sum(te_class == te_target) %>%
  divide_by(length(te_target)) %T>%
  {
    percent(.) %>%
      sprintf("Overall accuracy = %s\n", .) %>%
      cat()
  } ->
te_acc

## Overall accuracy=59.6%

sum(te_class != te_target) %>%
  divide_by(length(te_target)) %T>%
  {
    percent(.) %>%
      sprintf("Overall error = %s\n", .) %>%
      cat()
  } ->
te_err
```

```
## Overall error=40.4%

errorMatrix(te_target, te_class) %T>%
  print() ->
te_matrix

##        Predicted
## Actual no  yes Error
##     no  39 38.4  49.6
##     yes  2 20.5   9.1

te_rec <- (te_matrix[2,2]/(te_matrix[2,2]+te_matrix[2,1])) %T>%
  {percent(.) %>% sprintf("Recall = %s\n", .) %>% cat()}

## Recall=91.1%

te_pre <- (te_matrix[2,2]/(te_matrix[2,2]+te_matrix[1,2])) %T>%
  {percent(.) %>% sprintf("Precision = %s\n", .) %>% cat()}

## Precision=34.8%

te_fsc <- ((2 * te_pre * te_rec)/(te_rec + te_pre))  %T>%
  {sprintf("F-Score = %.3f\n", .) %>% cat()}

## F-Score=0.504

te_prob %>%
  prediction(te_target) %>%
  performance("auc") %>%
  attr("y.values") %>%
  .[[1]] %T>%
  {
    percent(.) %>%
    sprintf("Percentage area under the ROC curve = %s\n", .) %>%
    cat()
  } ->
te_auc

## Percentage area under the ROC curve=76.2%
```

Figures 7.7 and 7.8 plot the ROC curve and the risk chart, respectively, for the testing dataset.

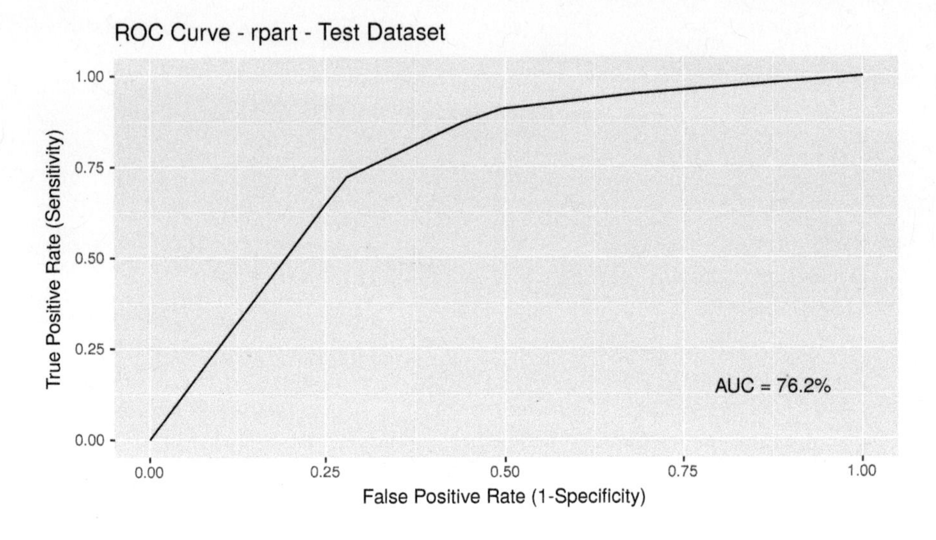

Figure 7.7: *An ROC curve for a decision tree on the testing dataset.*

```
te_prob %>%
  prediction(te_target) %>%
  performance("tpr", "fpr") ->
te_rates

data_frame(tpr=attr(te_rates, "y.values")[[1]],
           fpr=attr(te_rates, "x.values")[[1]]) %>%
  ggplot(aes(fpr, tpr)) +
  geom_line() +
  annotate("text", x=0.875, y=0.125, vjust=0,
           label=paste("AUC =", percent(te_auc))) +
  labs(title="ROC Curve - " %s+% mtype %s+% " - Test Dataset",
       x="False Positive Rate (1-Specificity)",
       y="True Positive Rate (Sensitivity)")
```

```
riskchart(te_prob, te_target, te_risk) +
  labs(title="Risk Chart - " %s+%
         mtype %s+%
         " - Validation Dataset") +
  theme(plot.title=element_text(size=14))
```

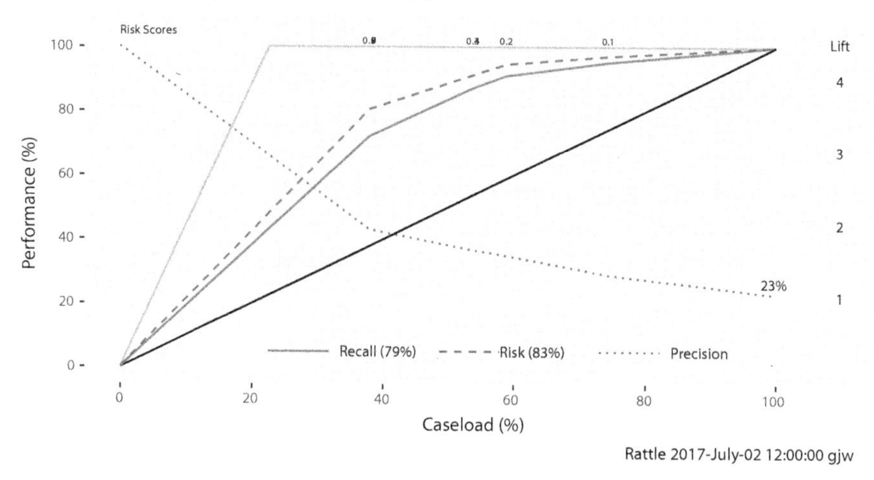

Figure 7.8: *A risk chart for the testing dataset.*

7.6 Comparison of Performance Measures

To wrap up our modelling we present, for a chosen model (`m_rp01`, for example), the various performance measures we have captured. Do note again that there are many other measures of performance that we can add into our modelling scripts but we choose these basics here as the starting point. Note also that we compute all of the performance measures for the three datasets based on the final model we built. The results can be seen in Table 7.1.

Table 7.1: *Performance measures for decision tree model.*

	Acc	Err	TN	FP	FN	TP	No	Yes	AUC	Re	Pr	Fs
Tr	88	12	75	3	9	13	4	41	86	59	82	68
Va	84	16	73	5	11	11	6	51	82	49	68	57
Te	83	17	72	5	11	11	7	50	83	50	68	57

7.7 Save the Model to File

Having completed our modelling (or even whilst we are iterating over model builds) we will want to save the model to file, together with a link to the dataset used. As with the data template we store our results into a binary R file that collects together the relevant variables in a compressed format that we can readily load into R at another time.

```
# Save model into a appropriate folder.

fpath <- "models"

# Timestamp for the dataset - this is the general approach.

mdate <- format(Sys.Date(), "%Y%m%d")

# Use a fixed timestamp to name our file for convenience.

mdate <- "20170702"

# Filename for the saved dataset.

mfile <- sprintf("%s_%s_%s.RData", dsname, mtype, mdate)

# Full path to the dataset.

fpath %>%
  file.path(mfile) %T>%
  print() ->
mrdata

## [1] "models/weatherAUS_rpart_20170702.RData"

# Ensure the path exists.

if (! dir.exists(fpath)) dir.create(fpath)

# Save relevant R objects to the binary RData file.

save(dsrdata,
     tr_class, tr_prob, tr_acc, tr_err, tr_auc,
```

```
    tr_matrix, tr_rec, tr_pre, tr_fsc,
    va_class, va_prob, va_acc, va_err, va_auc,
    va_matrix, va_rec, va_pre, va_fsc,
    te_class, te_prob,te_acc, te_err, te_auc,
    te_matrix, te_rec, te_pre, te_fsc,
    model, mtype, mdesc,
    file=mrdata)
```

Notice that we save the file name of the binary RData file containing the dataset we used to build the model (`dsrdata`) so that we can tie the original dataset to the model.

```
# Check the resulting file size in bytes.

file.size(mrdata) %>% comma()

## [1] "2,759,759"
```

The dataset, its metadata and the model are now stored in *models/weatherAUS_rpart_20170702.RData*. We can load this into R at a later time and replicate the generic process we have illustrated in this chapter.

```
load(mrdata) %>% print()

##  [1] "dsrdata"   "tr_class"  "tr_prob"   "tr_acc"
##  [5] "tr_err"    "tr_auc"    "tr_matrix" "tr_rec"
##  [9] "tr_pre"    "tr_fsc"    "va_class"  "va_prob"
## [13] "va_acc"    "va_err"    "va_auc"    "va_matrix"
## [17] "va_rec"    "va_pre"    "va_fsc"    "te_class"
## [21] "te_prob"   "te_acc"    "te_err"    "te_auc"
....
```

Note that by using generic variable names we can load different model files and perform common operations on them without changing the names within a script. However, do note that each time we load such a saved model file we overwrite any other variables of the same name.

7.8 A Template for Predictive Modelling

As we did in Chapter 3 throughout this chapter we have worked towards constructing a standard template for ourselves for a predictive modelling report. The template will provide a useful starting point for any predictive modelling.

A template based on this chapter for predictive modelling is available from `https://essentials.togaware.com`. As with the data template these templates will continue to be refined over time and will incorporate improvements and advanced techniques that go beyond what has been presented here.

7.9 Exercises

Exercise 7.1 Alternative Modellers

1. Replicate the model template as available from `https://essentials.togaware.com` using the **WeatherAUS** dataset.
2. Extend the instantiated template above by repeating the model build using alternative algorithms, including at least logistic regression and support vector machine.
3. Identify the knowledge, if any, gained from the newly built models.
4. Evaluate the performance of the new models and compare and contrast with the performance of the decision tree model.
5. Produce a report to share the new narrative and knowledge discovered.

Exercise 7.2 Binary Classification

1. Identify a dataset either from your own work experience, from amongst the many available freely on the Internet, or from an Internet competition site. The dataset would ideally include both numeric and categoric data and will need to have a suitable target variable for classification, and preferably for binary classification.

2. Using this dataset instantiate the data template available from `https://essentials.togaware.com` to prepare the dataset for predictive model building.

3. Using the dataset processed through the data template instantiate the model template available from `https://essentials.togaware.com` to build and evaluate a predictive model.

4. Produce a report to share the narrative, including the identification of the data source, the data wrangling performed, and the model building, together with an exploration of the knowledge you discover.

Exercise 7.3 Alternative Modellers

1. Extend the instantiated template developed above by repeating the model build using alternative algorithms, including logistic regression and support vector machine.

2. Identify the knowledge, if any, gained from the models built.

3. Include performance evaluation and compare to the decision tree model.

4. Produce a report to share the new narrative and knowledge discovered.

8

Ensemble of Predictive Models

In Chapter 7 we presented a process for building, viewing, and evaluating models. There we continued with the concept of a template, introduced in Chapter 3, that captures through the use of generic variables a repeatable process. We use the template as a starting point for any new task. In this chapter we make use of the model building template to illustrate the process with alternative machine learning algorithms.

A decision tree is one of the simplest machine learning models and is particularly popular because we can present the model to a domain expert and they will have a good chance of understanding the discovered knowledge. Other modelling approaches are less accessible in terms of explaining the knowledge they have discovered. Yet these other modelling approaches can deliver more accurate models.

A concept of combining multiple models for multiple inductive learning was found to result in increased predictive accuracy (Williams, 1987, 1988). This concept of ensembles of models remains today as a powerful mechanism and has grown to such an extent that almost all of the most accurate learners today are ensemble based in some form.

In this chapter we introduce two popular and effective machine learning algorithms which exhibit an ensemble approach. We particularly aim to repeat and reinforce the use of the template we have developed for model building.

Packages used in this chapter include Matrix (Bates and Maechler, 2017), ROCR (Sing *et al.*, 2015), dplyr (Wickham *et al.*, 2017a), ggplot2 (Wickham and Chang, 2016), magrittr (Bache and Wickham, 2014), randomForest (Breiman *et al.*, 2015), rattle (Williams, 2017), scales (Wickham, 2016), stringi (Gagolewski *et al.*,

2017), tibble (Müller and Wickham, 2017) and xgboost (Chen *et al.*, 2017).

```
# Load required packages from the local library into R.

library(Matrix)        # Wrangle: sparse.model.matrix().
library(ROCR)          # Evaluate: prediction().
library(dplyr)         # Wrangle: select()
library(ggplot2)       # Visualise: performance.
library(magrittr)      # Pipelines: %>% %T>% %<>%.
library(randomForest) # Model: randomForest() na.roughfix().
library(rattle)        # Evaluate: riskchart(), comcat().
library(scales)        # Formats: percent()
library(stringi)       # Strings: %s+%.
library(tibble)        # Data: as_data_frame()
library(xgboost)       # Models: extreme gradient boosting.
```

8.1 Loading the Dataset

The starting point for model building is to load the dataset we carefully crafted and saved in Chapter 3.

```
# Build the filename used to previously store the data.

fpath  <- "data"
dsname <- "weatherAUS"
dsdate <- "_20170702"
dsfile <- dsname %s+% dsdate %s+% ".RData"

fpath %>%
  file.path(dsfile) %T>%
  print() ->
dsrdata

## [1] "data/weatherAUS_20170702.RData"

# Load the R objects from file and list them.

load(dsrdata) %>% print()

##  [1] "ds"          "dsname"      "dspath"      "dsdate"
##  [5] "nobs"        "vars"        "target"      "risk"
```

```
##  [9] "id"         "ignore"     "omit"       "inputi"
## [13] "inputs"     "numi"       "numc"       "cati"
## [17] "catc"       "form"       "seed"       "train"
## [21] "validate"   "test"       "tr_target"  "tr_risk"
....
```

A quick review should tell us that the dataset appears to load just fine and matches our expectations for modelling.

8.2 Random Forest

A random forest is an example of an ensemble model that can significantly improve the classification accuracy over a decision tree. The random forest algorithm samples the training dataset multiple times, each being used to build a decision tree. Subsets of the variables within the training dataset are also randomly chosen in building each of the decision trees.

The random forest algorithm is implemented in R as `randomForest::randomForest()`. Below we build a default random forest, with 500 trees, and then deploy our template to evaluate the model performance.

We note this particular implementation of the random forest algorithm does not handle missing values for any of the variables in any observation and so we use `randomForest::na.roughfix()` as the value for the `na.action=` to impute missing values (which is not always appropriate). We also add an option to ask for measures of the `importance=` of the variables to be calculated and returned by the function.

```
# Train a random forest model.

m_rf <- randomForest(form,
                     data=ds[train, vars],
                     na.action=na.roughfix,
                     importance=TRUE)
```

Having built our model we copy it to our template generic variables and the remainder of the evaluation will be the same as for any model.

```
# Initialise generic variables.

model <- m_rf
mtype <- "randomForest"
mdesc <- "random forest"

# Basic model structure.

model

##
## Call:
##  randomForest(formula=form, data=ds[train, vars], impo...
##                Type of random forest: classification
##                      Number of trees: 500
## No. of variables tried at each split: 4
##
##         OOB estimate of  error rate: 14.44%
## Confusion matrix:
##        no   yes class.error
## no  69704  3395  0.04644386
## yes 10232 11012  0.48164188
```

We now have 500 decision trees forming the ensemble model. It is generally not possible to visualise all 500 trees. To get some insight into the discovered knowledge behind the model we can review the variable importance plots using `rattle::ggVarImp()`. Two measures are generated, one using a gini calculation and the other based on accuracy. The relative importance of a variable with respect to each of the predicted classes is also provided. See Figure 8.1.

```
# Review variable importance.

ggVarImp(model, log=TRUE)
```

Now we are ready to review the performance of the model. We do so using the validation dataset.

```
# Predict on the validation dataset.

model %>%
```

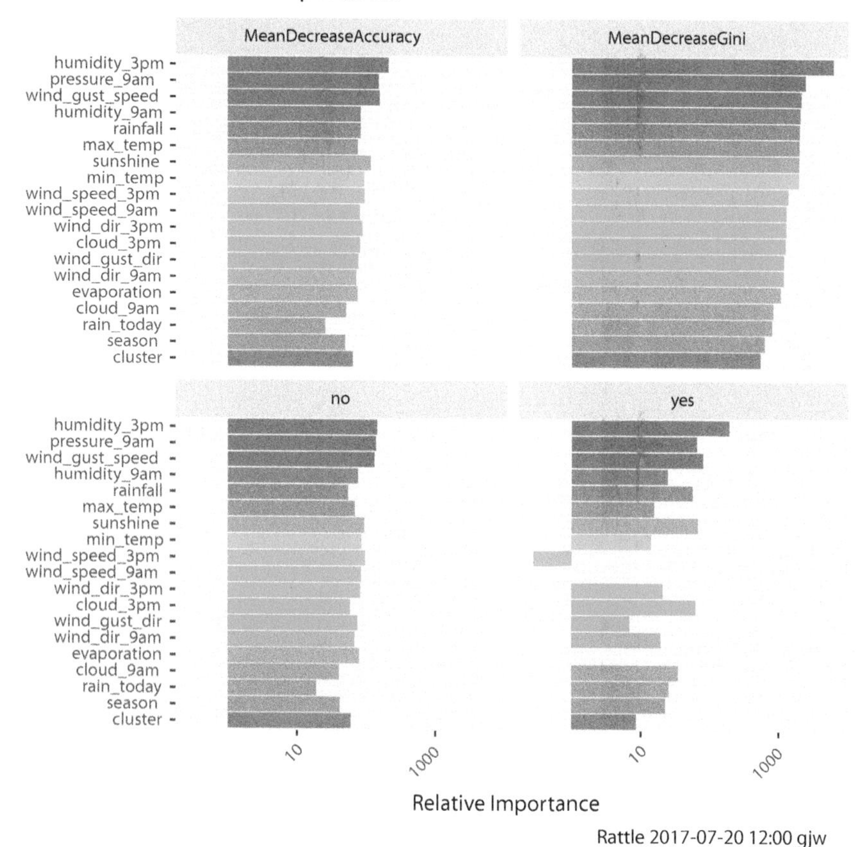

Figure 8.1: *Random forest variable importance.*

```
  predict(newdata=ds[validate, vars], type="prob") %>%
  .[,2] %>%
  set_names(NULL) %>%
  round(2) %T>%
  {head(., 20) %>% print()} ->
va_prob

## [1] 0.88 0.02 0.05 0.11 0.04   NA 0.04   NA   NA   NA   NA
## [12] 0.09   NA 0.24 0.13 0.48   NA   NA   NA 0.23

model %>%
  predict(newdata=ds[validate, vars], type="response") %>%
  set_names(NULL) %T>%
```

```
  {head(., 20) %>% print()} ->
va_class

## [1] yes  no   no   no   no   <NA> no   <NA> <NA> <NA> <NA>
## [12] no   <NA> no   no   no   <NA> <NA> <NA> no
## Levels: no yes
```

Notice that predictions are not made for observations with any missing values for any of the input variables and thus the result is `NA`. Consequently our formula for calculating the accuracy and error needs to be modified to cope with this by including `na.rm=` and checking for missing values. Indeed we would update our template appropriately to handle either scenarios (either complete data or including missing data).

```
sum(va_class == va_target, na.rm=TRUE) %>%
  divide_by(va_class %>% is.na() %>% not() %>% sum()) %T>%
  {
    percent(.) %>%
    sprintf("Overall accuracy = %s\n", .) %>%
      cat()
  } ->
va_acc

## Overall accuracy=86.7%

sum(va_class != va_target, na.rm=TRUE) %>%
  divide_by(va_class %>% is.na() %>% not() %>% sum()) %T>%
  {
    percent(.) %>%
    sprintf("Overall error = %s\n", .) %>%
      cat()
  } ->
va_err

## Overall error=13.3%
```

The overall accuracy is in line with that of the decision tree model of Chapter 7.

```
# Count of prediction versus actual as a confusion matrix.

errorMatrix(va_target, va_class, count=TRUE)

##        Predicted
## Actual   no yes Error
##     no 6185 300   4.6
##    yes  795 933  46.0

# Comparison as percentages of all observations.

errorMatrix(va_target, va_class) %T>%
  print() ->
va_matrix

##        Predicted
## Actual   no  yes Error
##     no 75.3  3.7   4.6
##    yes  9.7 11.4  46.0

va_matrix %>%
  diag() %>%
  sum(na.rm=TRUE) %>%
  subtract(100, .) %>%
  sprintf("Overall error percentage = %s%%\n", .) %>%
  cat()

## Overall error percentage=13.3%

va_matrix[,"Error"] %>%
  mean(na.rm=TRUE) %>%
  sprintf("Averaged class error percentage = %s%%\n", .) %>%
  cat()

## Averaged class error percentage=25.3%

# AUC.

va_prob %>%
  prediction(va_target) %>%
  performance("auc") %>%
  attr("y.values") %>%
  .[[1]] %T>%
  {
```

```
    percent(.) %>%
    sprintf("Percentage area under the ROC curve = %s\n", .) %>%
    cat()
  } ->
va_auc

## Percentage area under the ROC curve=90.1%

# Recall, precision, and F-score.

va_rec <- (va_matrix[2,2]/(va_matrix[2,2]+va_matrix[2,1])) %T>%
  {percent(.) %>% sprintf("Recall = %s\n", .) %>% cat()}

## Recall=54%

va_pre <- (va_matrix[2,2]/(va_matrix[2,2]+va_matrix[1,2])) %T>%
  {percent(.) %>% sprintf("Precision = %s\n", .) %>% cat()}

## Precision=75.5%

va_fsc <- ((2 * va_pre * va_rec)/(va_rec + va_pre))  %T>%
  {sprintf("F-Score = %.3f\n", .) %>% cat()}

## F-Score=0.630

# Rates for ROC curve.

va_prob %>%
  prediction(va_target) %>%
  performance("tpr", "fpr") %>%
  print() ->
va_rates

## An object of class "performance"
## Slot "x.name":
## [1] "False positive rate"
##
## Slot "y.name":
## [1] "True positive rate"
....
```

Figures 8.2 and 8.3 illustrate that out of the box (without any

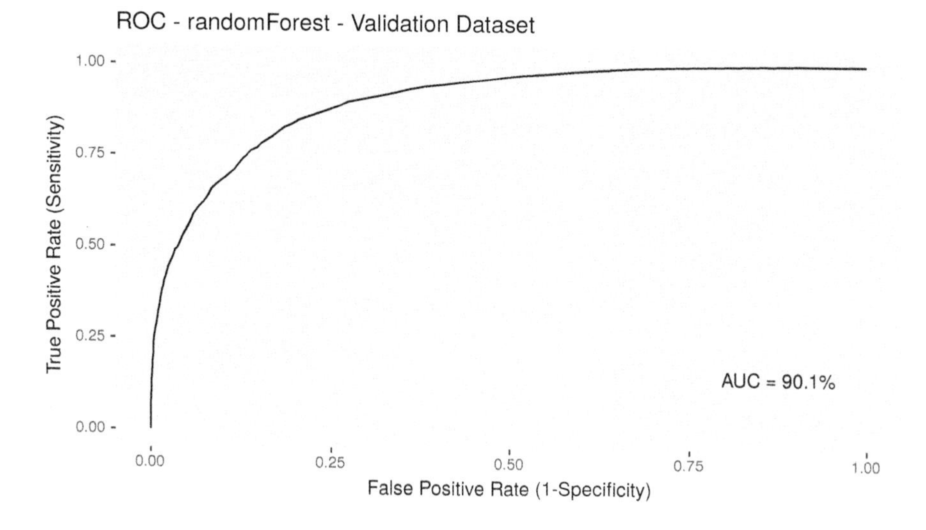

Figure 8.2: *ROC for random forest over validation dataset.*

tuning of the algorithm) the random forest model performs quite well.

```
data_frame(tpr=attr(va_rates, "y.values")[[1]],
           fpr=attr(va_rates, "x.values")[[1]]) %>%
  ggplot(aes(fpr, tpr)) +
  geom_line() +
  annotate("text", x=0.875, y=0.125, vjust=0,
           label=paste("AUC =", percent(va_auc))) +
  labs(title="ROC - " %s+% mtype %s+% " - Validation Dataset",
       x="False Positive Rate (1-Specificity)",
       y="True Positive Rate (Sensitivity)")
```

```
riskchart(va_prob, va_target, va_risk) +
  labs(title="Risk Chart - " %s+%
         mtype %s+%
         " - Validation Dataset") +
  theme(plot.title=element_text(size=14))
```

It is instructive to review the error matrix, the ROC curve, and the risk chart using the training dataset. The basic evaluation below, presenting just the output from the code rather than the

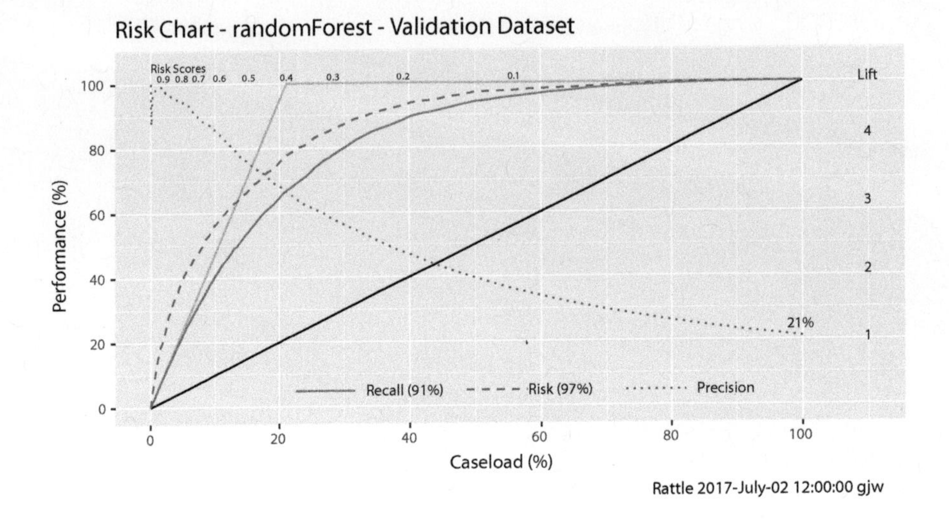

Figure 8.3: *Risk chart random forest validation dataset.*

code itself, shows that we have a perfect model—the model perfectly predicts the observations it was trained on.

```
## Performance evaluation random forest training dataset.
## Overall accuracy=100%
## Overall error=0%
##         Predicted
## Actual      no   yes Error
##     no  29879     0     0
##     yes     0  8475     0
##         Predicted
## Actual    no  yes Error
##     no  77.9  0.0     0
##     yes  0.0 22.1     0
## Recall=100%
## Precision=100%
## F-Score=1.000
## Percentage area under the ROC curve=100%
```

Figures 8.4 and 8.5 further confirm that the random forest model is a perfect fit for the training dataset. As we noted in Chapter 7 though, the performance on the training set is an optimistic estimate of how well the model will perform in general.

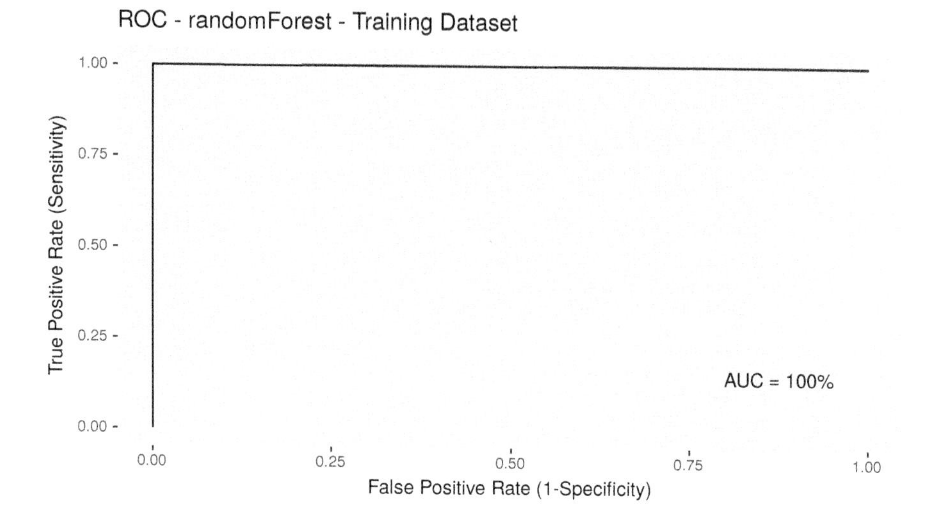

Figure 8.4: *Random forest ROC over training dataset.*

The performance on the validation dataset above indicates a more realistic estimate of how well the model will perform in general.

We similarly compute performance measures for the test dataset to produce the summary presented in Table 8.1.

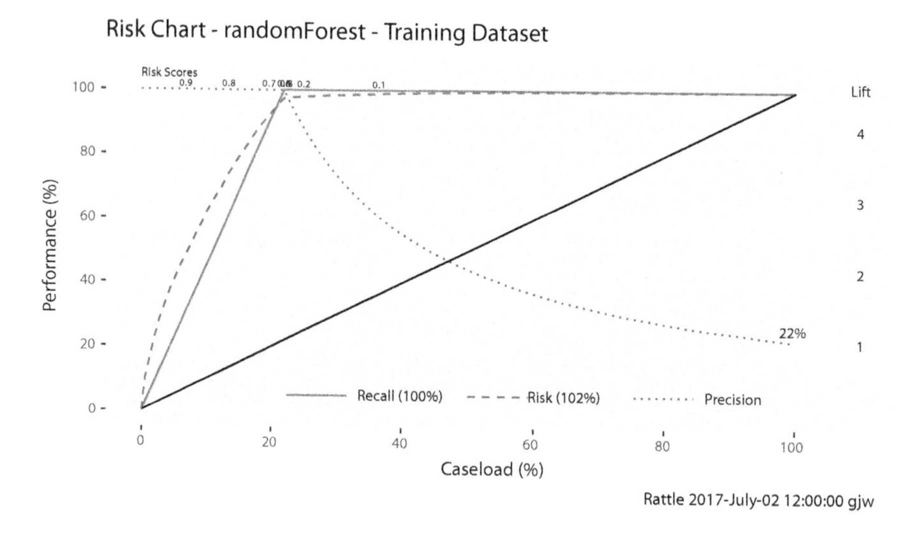

Figure 8.5: *Random forest risk chart over training dataset.*

```
## Performance evaluation random forest test dataset.
## Overall accuracy=86%
## Overall error=14%
##        Predicted
## Actual    no  yes Error
##     no  6027  312   4.9
##     yes  838 1013  45.3
##        Predicted
## Actual    no  yes Error
##     no  73.6  3.8   4.9
##     yes 10.2 12.4  45.3
## Recall=54.9%
## Precision=76.5%
## F-Score=0.639
## Percentage area under the ROC curve=89.8%
```

Table 8.1: *Performance measures for the random forest model.*

	Acc	Err	TN	FP	FN	TP	No	Yes	AUC	Re	Pr	Fs
Tr	100	0	78	0	0	22	0	0	100	100	100	100
Va	87	13	75	4	10	11	5	46	90	54	76	63
Te	86	14	74	4	10	12	5	45	90	55	76	64

As always it is a good idea to now save the performance data and the model for future reference.

```
# Save model into an appropriate folder.

fpath <- "models"

# Timestamp for the dataset - this is the general approach.

mdate <- format(Sys.Date(), "%Y%m%d")

# Use a fixed timestamp to name our file for convenience.

mdate <- "20170702"

# Filename for the saved dataset.

mfile <- sprintf("%s_%s_%s.RData", dsname, mtype, mdate)
```

```
# Full path to the dataset.

fpath %>%
  file.path(mfile) %T>%
  print() ->
mrdata

## [1] "models/weatherAUS_randomForest_20170702.RData"

# Ensure the path exists.

if (! dir.exists(fpath)) dir.create(fpath)

# Save relevant R objects to the binary RData file.

save(dsrdata,
     tr_class, tr_prob, tr_acc, tr_err, tr_auc,
     tr_matrix, tr_rec, tr_pre, tr_fsc,
     va_class, va_prob, va_acc, va_err, va_auc,
     va_matrix, va_rec, va_pre, va_fsc,
     te_class, te_prob,te_acc, te_err, te_auc,
     te_matrix, te_rec, te_pre, te_fsc,
     model, mtype, mdesc,
     file=mrdata)

# Check the resulting file size in bytes.

file.size(mrdata) %>% comma()

## [1] "61,781,151"
```

Notice that this file size is considerably larger than that for the decision tree in Chapter 7. A random forest model consists of many decision trees.

8.3 Extreme Gradient Boosting

A very popular algorithm for building an ensemble of decision trees in parallel uses an approach known as extreme gradient boosting. It is implemented and made available in R as `xgboost::xgboost()`.

The concept of the algorithm is to build a series of decision trees, each attempting to boost its performance on the misclassified observations of the previous decision tree. We then have an ensemble of trees which each predict and are combined through a weighted aggregation to calculate the final prediction.

The algorithm accepts only numeric data and so categoric variables need to be converted to numeric. We do this using an approach where each value of a categoric variable is turned into a variable itself. Then for any observation these variables are all 0 except for that which corresponds to the value of the original variable which is 1.

We can use `Matrix::sparse.model.matrix()` to perform this operation and build a sparse dataset structure suitable for `xgboost::xgboost()`. In the following code we especially identify the target variable in the formula to ignore it in the resulting sparse matrix. Notice also that for illustrative purposes we impute missing values in our dataset using `randomForest::na.roughfix()`.

```
formula(target %s+% "~ .-1") %>%
  sparse.model.matrix(data=ds[vars] %>% na.roughfix()) %T>%
  {dim(.) %>% print()} %T>%
  {head(.) %>% print()} ->
sds

## [1] 134776     67
## 6 x 67 sparse Matrix of class "dgCMatrix"

##     [[ suppressing 67 column names 'min_temp', 'max_temp',
'rainfall' ... ]]

##
## 1 13.4 22.9 0.6 4.8 8.5 . . . . . . . . . . . . . 1 . . . 44
## 2  7.4 25.1 .   4.8 8.5 . . . . . . . . . . . . . . 1 . . 44
## 3 12.9 25.7 .   4.8 8.5 . . . . . . . . . . . . 1 . . . . 46
## 4  9.2 28.0 .   4.8 8.5 . . 1 . . . . . . . . . . . . . . 24
## 5 17.5 32.3 1.0 4.8 8.5 . . . . . . . . . . . . . 1 . . . 41
....
```

Notice the dimensions of the resulting sparse matrix and in particular the 67 columns. Multiple new variables have been introduced to replace each of the categoric variables.

We also need a vector to record the values of the target variable. Note that our dataset is an extended *tbl* data frame and so we need to extract the column into a vector using `base::unlist()`. The target variable is a binary variable (having just two values) and so we turn this into a Boolean.

```
ds[target] %>%
  unlist(use.names=FALSE) %>%
  equals("yes") %T>%
  {head(., 20) %>% print()} ->
label
```

```
##  [1] FALSE FALSE FALSE FALSE FALSE FALSE FALSE FALSE  TRUE
## [10] FALSE  TRUE  TRUE  TRUE FALSE FALSE  TRUE  TRUE FALSE
## [19] FALSE FALSE
```

We are now ready to build a model using the extreme gradient boosting algorithm. We set the maximum number of iterations (i.e., number of trees) as `nrounds=100` and note that we wish to build a binary logistic model.

```
# Train an extreme gradient boosting model.

m_xg <- xgboost(data=sds[train,],
                label=label[train],
                nrounds=100,
                print_every_n=15,
                objective="binary:logistic")
```

```
## [1] train-error:0.157680
## [16] train-error:0.142194
## [31] train-error:0.136481
## [46] train-error:0.129835
## [61] train-error:0.124832
## [76] train-error:0.118567
## [91] train-error:0.114105
## [100] train-error:0.111784
```

For `print_every_n=15` iterations the decreasing error rate is reported.

As usual we copy the model to our template generic variables to minimally change any of the following evaluation code.

```
# Initialise generic variables.

model <- m_xg
mtype <- "xgboost"
mdesc <- "extreme gradient boosting"

# Basic model information.

model

## ##### xgb.Booster
## raw: 363.8 Kb
## call:
##   xgb.train(params=params, data=dtrain, nrounds=nrou...
##     watchlist=watchlist, verbose=verbose, print_every_...
##     early_stopping_rounds=early_stopping_rounds, maximiz...
##     save_period=save_period, save_name=save_name, xgb_...
##     callbacks=callbacks, objective="binary:logistic")
## params (as set within xgb.train):
##   objective="binary:logistic", silent="1"
## xgb.attributes:
##   niter
## callbacks:
##   cb.print.evaluation(period=print_every_n)
##   cb.evaluation.log()
##   cb.save.model(save_period=save_period, save_name=sav...
## niter: 100
## evaluation_log:
##     iter train_error
##        1    0.157680
##        2    0.155221
## ---
##       99    0.112059
##      100    0.111784
```

With 100 trees it is not generally feasible to print every tree to review the discovered knowledge. Indeed, even in studying every tree the complexity of how the decisions from each tree are combined makes it rather difficult to finely understand the details. And so the variable importance plot will again provide insight into the discovered knowledge. The plot reports on what the algorithm identifies as the relative importance of the input variables. For xgboost the variable names are not stored with the model itself so

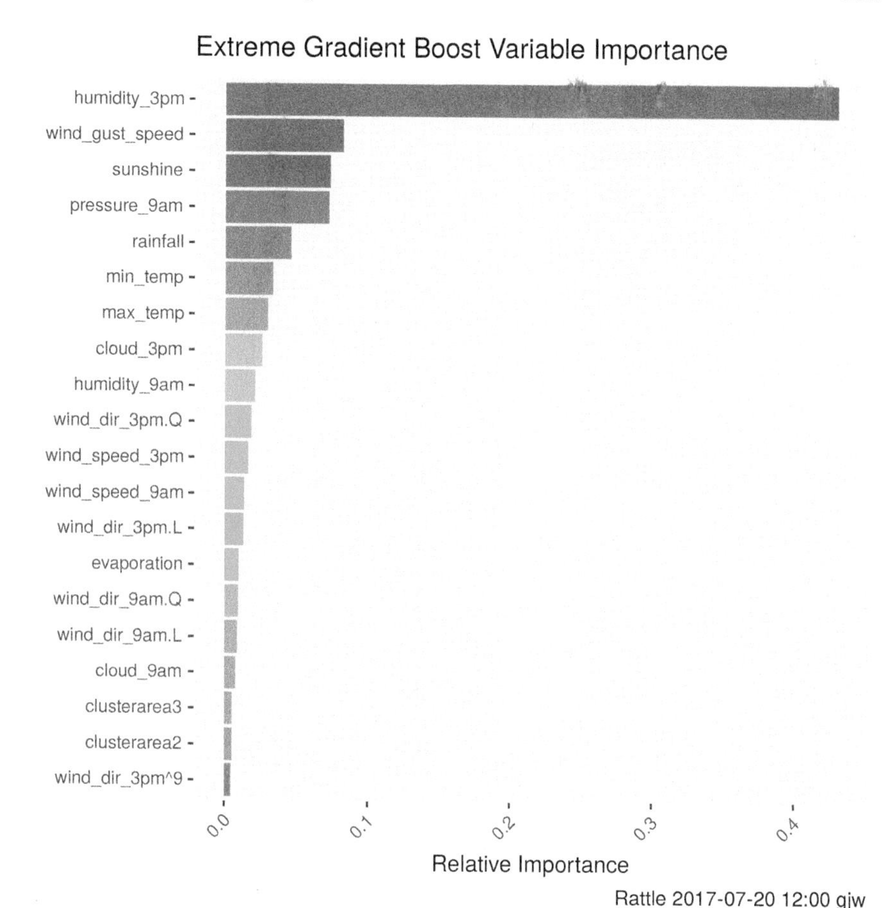

Figure 8.6: *Extreme gradient boosting variable importance.*

we need to pass them along to `rattle::ggVarImp()`. Also since there are a large number of variables (having turned each categoric value into a variable) we only display the top 20 variables. Figure 8.6 contains a similar analysis of variable importance as we have previously seen.

```
# Review variable importance.

ggVarImp(model, feature_names=colnames(sds), n=20)
```

The `stats::predict()` function for the xgboost model returns the probabilities. In the following code block we record the probabilities and then convert them to a decision with a 0.50 threshold.

```
# Predict on the validate dataset.

model %>%
  predict(newdata=sds[validate,]) %>%
  set_names(NULL) %>%
  round(2) %T>%
  {head(., 20) %>% print()} ->
va_prob

##  [1] 0.94 0.03 0.03 0.07 0.03 0.26 0.02 0.68 0.32 0.56 0.22
## [12] 0.06 0.02 0.16 0.11 0.56 0.20 0.06 0.06 0.22

va_prob %>%
  is_greater_than(0.5) %>%
  ifelse("yes", "no") %T>%
  {head(., 20) %>% print()} ->
va_class

##  [1] "yes" "no"  "no"  "no"  "no"  "no"  "no"  "yes" "no"
## [10] "yes" "no"  "no"  "no"  "no"  "no"  "yes" "no"  "no"
## [19] "no"  "no"
```

The remainder of our evaluation is the same as per the generic template.

```
sum(va_class == va_target, na.rm=TRUE) %>%
  divido_by(va class %>% is.na() %>% not() %>% sum()) %T>%
  {
    percent(.) %>%
    sprintf("Overall accuracy = %s\n", .) %>%
      cat()
  } ->
va_acc

## Overall accuracy=85.3%

sum(va_class != va_target, na.rm=TRUE) %>%
  divide_by(va_class %>% is.na() %>% not() %>% sum()) %T>%
  {
    percent(.) %>%
    sprintf("Overall error = %s\n", .) %>%
```

```
    cat()
  } ->
va_err

## Overall error=14.7%

# Count of prediction versus actual as a confusion matrix.

errorMatrix(va_target, va_class, count=TRUE)

##          Predicted
## Actual      no  yes Error
##     no   14846  894   5.7
##     yes   2074 2402  46.3

# Comparison as percentages of all observations.

errorMatrix(va_target, va_class) %T>%
  print() ->
va_matrix

##          Predicted
## Actual     no  yes Error
##     no   73.4  4.4   5.7
##     yes  10.3 11.9  46.3

va_matrix %>%
  diag() %>%
  sum(na.rm=TRUE) %>%
  subtract(100, .) %>%
  sprintf("Overall error percentage = %s%%\n", .) %>%
  cat()

## Overall error percentage=14.7%

va_matrix[,"Error"] %>%
  mean(na.rm=TRUE) %>%
  sprintf("Averaged class error percentage = %s%%\n", .) %>%
  cat()

## Averaged class error percentage=26%
```

```
# AUC.

va_prob %>%
  prediction(va_target) %>%
  performance("auc") %>%
  attr("y.values") %>%
  .[[1]] %T>%
  {
    percent(.) %>%
    sprintf("Percentage area under the ROC curve = %s\n", .) %>%
    cat()
  } ->
va_auc

## Percentage area under the ROC curve=88.2%

# Recall, precision, and F-score.

va_rec <- (va_matrix[2,2]/(va_matrix[2,2]+va_matrix[2,1])) %T>%
  {percent(.) %>% sprintf("Recall = %s\n", .) %>% cat()}

## Recall=53.6%

va_pre <- (va_matrix[2,2]/(va_matrix[2,2]+va_matrix[1,2])) %T>%
  {percent(.) %>% sprintf("Precision = %s\n", .) %>% cat()}

## Precision=73%

va_fsc <- ((2 * va_pre * va_rec)/(va_rec + va_pre))  %T>%
  {sprintf("F-Score = %.3f\n", .) %>% cat()}

## F-Score=0.618

# Rates for ROC curve.

va_prob %>%
  prediction(va_target) %>%
  performance("tpr", "fpr") %>%
  print() ->
va_rates

## An object of class "performance"
## Slot "x.name":
## [1] "False positive rate"
##
## Slot "y.name":
## [1] "True positive rate"
....
```

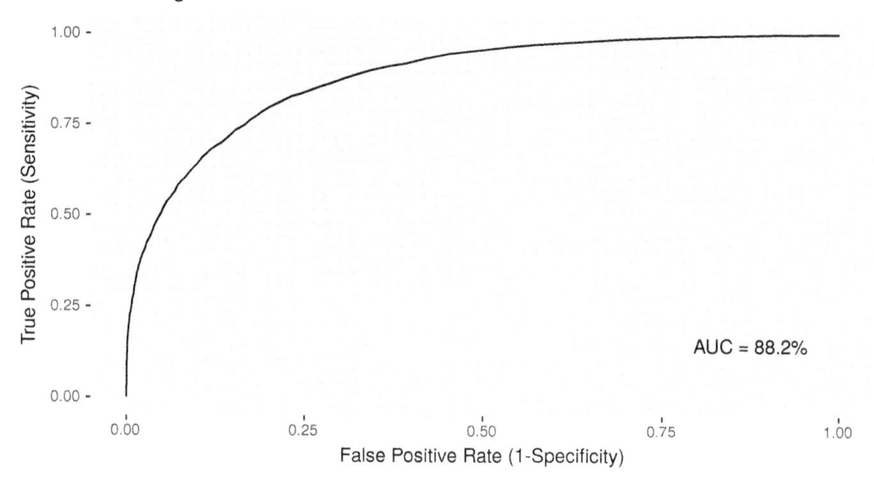

Figure 8.7: *ROC for extreme gradient boosting.*

Figures 8.7 and 8.8 illustrate that out of the box (without any tuning of the algorithm) the model performs quite well.

```
data_frame(tpr=attr(va_rates, "y.values")[[1]],
           fpr=attr(va_rates, "x.values")[[1]]) %>%
  ggplot(aes(fpr, tpr)) +
  geom_line() +
  annotate("text", x=0.875, y=0.125, vjust=0,
           label=paste("AUC =", percent(va_auc))) +
  labs(title="ROC - " %s+% mtype %s+% " - Validation Dataset",
       x="False Positive Rate (1-Specificity)",
       y="True Positive Rate (Sensitivity)")
```

```
riskchart(va_prob, va_target, va_risk) +
  labs(title="Risk Chart - " %s+%
         mtype %s+%
         " - Validation Dataset") +
  theme(plot.title=element_text(size=14))
```

Overall the performance is on par with the random forest. It should be noted that we have not done any model tuning for either the random forest or the extreme gradient boosting. Typically this

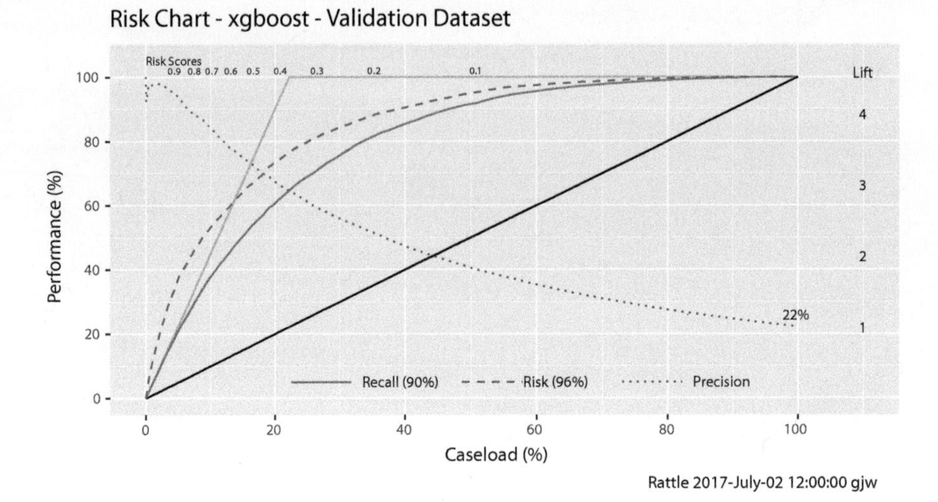

Figure 8.8: *Risk chart for extreme gradient boosting.*

is our starting point for now tuning the algorithm to target the best performing model.

As per our template we now repeat the performance evaluation on both the training and the testing datasets. Recall that the training dataset will provide an optimistic estimate and the testing dataset is used once we have finished tuning the model build to gauge how the model will perform with new unseen data.

```
## Performance evaluation extreme gradient boosting training ...
## Overall accuracy=88.8%
## Overall error=11.2%
##        Predicted
## Actual     no   yes Error
##    no  70599  2500   3.4
##    yes  8074 13170  38.0
##        Predicted
## Actual    no  yes Error
##    no  74.8  2.6   3.4
##    yes  8.6 14.0  38.0
## Recall=61.9%
## Precision=84.3%
## F-Score=0.714
## Percentage area under the ROC curve=93.2%
```

```
## Performance evaluation extreme gradient boosting test data...
## Overall accuracy=85.2%
## Overall error=14.8%
##        Predicted
## Actual    no  yes Error
##     no 14773  881   5.6
##     yes  2103 2460  46.1
##        Predicted
## Actual   no  yes Error
##     no 73.1  4.4   5.6
##     yes 10.4 12.2  46.1
## Recall=54%
## Precision=73.5%
## F-Score=0.622
## Percentage area under the ROC curve=88.2%
```

Table 8.2: *Performance measures extreme gradient boosting*

	Acc	Err	TN	FP	FN	TP	No	Yes	AUC	Re	Pr	Fs
Tr	89	11	75	3	9	14	3	38	93	62	84	71
Va	85	15	73	4	10	12	6	46	88	54	73	62
Te	85	15	73	4	10	12	6	46	88	54	74	62

As always it is a good idea to now save the performance data and model for future reference.

```
# Save model into an appropriate folder.

fpath <- "models"

# Timestamp for the dataset - this is the general approach.

mdate <- format(Sys.Date(), "%Y%m%d")

# Use a fixed timestamp to name our file for convenience.

mdate <- "20170702"

# Filename for the saved dataset.

mfile <- sprintf("%s_%s_%s.RData", dsname, mtype, mdate)
```

```
# Full path to the dataset.

fpath %>%
  file.path(mfile) %T>%
  print() ->
mrdata

## [1] "models/weatherAUS_xgboost_20170702.RData"

# Ensure the path exists.

if (! dir.exists(fpath)) dir.create(fpath)

# Save relevant R objects to the binary RData file.

save(dsrdata,
     tr_class, tr_prob, tr_acc, tr_err, tr_auc,
     tr_matrix, tr_rec, tr_pre, tr_fsc,
     va_class, va_prob, va_acc, va_err, va_auc,
     va_matrix, va_rec, va_pre, va_fsc,
     te_class, te_prob,te_acc, te_err, te_auc,
     te_matrix, te_rec, te_pre, te_fsc,
     sds, model, mtype, mdesc,
     file=mrdata)

# Check the resulting file size in bytes.

file.size(mrdata) %>% comma()

## [1] "18,918,360"
```

Notice the file size. We are also saving the sparse matrix version of the dataset along with the model and performance parameters.

8.4 Exercises

Exercise 8.1 Ensemble Modellers

1. Using the dataset identified for the exercises in Chapter 7 instantiate the model template available from `https://essentials.togaware.com` to build and evaluate a predictive model.

2. Produce a report to share the narrative of the model build, together with an exploration and discussion of the knowledge you discover.

Exercise 8.2 Deep Neural Networks

Deep neural networks are essentially an ensemble of a massive number of numeric computational nodes connected to flow data through a network to make a prediction.

1. For the same dataset identified for the exercises in Chapter 7 instantiate the model template to build and evaluate a predictive model using a deep neural network algorithm.

2. Produce a report to share the narrative of the model build, together with an exploration and discussion of any knowledge you discover.

9

Writing Functions in R

Over the preceding chapters we have repeatedly run a series of codes to perform an evaluation of the models we have built. The same code is repeated identically each time. The DRY (don't repeat yourself) principle of software engineering encourages us to recognise this as bad practise. Instead, all programming languages provide mechanisms to avoid this and in the end to produce more readily transparent, reproducible and efficient code.

In R as in many programming languages we can define our own functions. Chapter 2 introduced the concept of functions and we have seen small helper functions defined in previous chapters. Here we will create a function for ourselves. Our function will perform the evaluations presented in the preceding chapters.

Packages used in this chapter include ROCR (Sing *et al.*, 2015), magrittr (Bache and Wickham, 2014), randomForest (Breiman *et al.*, 2015), rattle (Williams, 2017), scales (Wickham, 2016) and stringi (Gagolewski *et al.*, 2017).

```
# Load required packages from the local library into R.

library(ROCR)         # Evaluate: prediction().
library(magrittr)     # Pipelines: %>% %T>% %<>%.
library(randomForest) # Model: randomForest() na.roughfix().
library(rattle)       # Data: weatherAUS.
library(scales)       # Formats: percent()
library(stringi)      # Strings: %s+%.
```

9.1 Model Evaluation

We will use the random forest model to illustrate and begin by loading the data and the model. We won't repeat the code here but report the output.

```
## [1] "data/weatherAUS_20170702.RData"
##  [1] "ds"          "dsname"      "dspath"      "dsdate"
##  [5] "nobs"        "vars"        "target"      "risk"
##  [9] "id"          "ignore"      "omit"        "inputi"
## [13] "inputs"      "numi"        "numc"        "cati"
## [17] "catc"        "form"        "seed"        "train"
## [21] "validate"    "test"        "tr_target"   "tr_risk"
....
## [1] "models/weatherAUS_randomForest_20170702.RData"
##  [1] "dsrdata"     "tr_class"    "tr_prob"     "tr_acc"
##  [5] "tr_err"      "tr_auc"      "tr_matrix"   "tr_rec"
##  [9] "tr_pre"      "tr_fsc"      "va_class"    "va_prob"
## [13] "va_acc"      "va_err"      "va_auc"      "va_matrix"
## [17] "va_rec"      "va_pre"      "va_fsc"      "te_class"
## [21] "te_prob"     "te_acc"      "te_err"      "te_auc"
....
```

We can then perform an evaluation of the performance of the model. We repeat below the evaluation using the test dataset, showing only the output since we have previously seen the code.

```
## Performance evaluation random forest test dataset.
## Overall accuracy=86%
## Overall error=14%
##       Predicted
## Actual    no  yes Error
##    no   6028  311   4.9
##    yes   839 1012  45.3
##       Predicted
## Actual    no  yes Error
##    no   73.6  3.8   4.9
##    yes  10.2 12.4  45.3
## Recall=54.9%
## Precision=76.5%
## F-Score=0.639
## Percentage area under the ROC curve=89.8%
```

In reviewing the previous code block we will note that the code produces both output and stores results into variables. We will want to do the same with the function that we write.

9.2 Creating a Function

As was emphasised in Chapter 2 in introducing the concept of building a pipeline, or building a plot, a function can generally be incrementally built as we need it. We often begin with a simple function and grow it from there. With experience we will more carefully design our functions before we start writing them.

9.2.1 Initial Messaging

We begin here with a simple function to initially print a message as in our above script.

```
perf <- function()
{
  # Provide informative introduction.

  sprintf("Performance evaluation %s test dataset.\n",mdesc) %>%
  cat()
}
```

Once we have defined the function we can call it directly.

```
perf()
```

```
## Performance evaluation random forest test dataset.
```

Immediately we notice that the printed title refers to just the random forest model and the test dataset. Both of these vary each time we conduct a performance evaluation. They are parameters of the performance evaluation and so we need a mechanism to change them each time the function is called.

In fact we might notice that the variable `mdesc` is already being used within the `base::sprintf()` and so we can simply change the value of that variable and call the function again.

```
mdesc <- "some fancy predictive model"
perf()

## Performance evaluation some fancy predictive model test da...
```

That works! Noting that the dataset is also changing from one call to `perf()` to the next we might create a variable for that.

```
perf <- function()
{
  # Provide informative introduction.

  sprintf("Performance evaluation %s %s dataset.\n",
          mdesc, dstype) %>%
  cat()
}

mdesc  <- "random forest"
dstype <- "test"
perf()

## Performance evaluation random forest test dataset.
```

So far so good, though we seem to be making changes to variables (`mdesc` and `dstype`) that could have impact on other parts of our code later on that also use these same variables. These are referred to as global variables since they are globally accessible—across all of our code. We can also modify the variables within the function and they will have any such new values globally. We have learnt over many decades of experience that using global variables, and especially changing the values of global variables within functions, can make our code difficult to understand and can lead to serious bugs.

Instead of relying on global variables a function allows us to pass variables in as arguments to the function call. This is the usual mechanism we use to pass data to functions.

```
perf <- function(mdesc, dstype)
{
  # Provide informative introduction.
```

```
  sprintf("Performance evaluation %s %s dataset.\n",
          mdesc, dstype) %>%
  cat()
}

perf("my favourite predictive model", "specially tuned")

## Performance evaluation my favourite predictive model speci...
```

Notice we can provide the values as part of the function call as string literals.

We can alternatively use previously defined variables, even of the same name. In this case the variables named within the function will have no relationship to these named variables passed as arguments except the value will be passed into the function and assigned to the local version of the variable. Their values will be provided to the variables internal to the function. Any changes made to the variables within the function will remain within the function. The global variables wil not be affected.

```
perf(mdesc, dstype)

## Performance evaluation random forest test dataset.
```

We may even be using very differently named variables to pass the data into the function.

```
model_description <- "random forest"
dataset_type <- "training dataset"
perf(model_description, dataset_type)

## Performance evaluation random forest training dataset data...
```

Finally, perhaps we now prefer to "improve" the messaging. A single change to our function will effect the change for every instance of its use.

```
perf <- function(mdesc, dstype)
{
  # Provide informative introduction.
```

```
  "Performance Evaluation\n" %s+%
    "======================\n\n" %s+%
    "Model:    " %s+% mdesc %s+% "\n" %s+%
    "Dataset: " %s+% dstype %s+% "\n" %>%
    cat()
}
perf(mdesc, dstype)

## Performance Evaluation
## ======================
##
## Model:   random forest
## Dataset: test
```

Imagine the effort required if we wanted to use this new format and had to go back through the code in the previous chapters where we have presented the model performance. We would need to make the change at least a dozen times! If we had been using this function all along, then no change would have been required there at all—only the function definition would need to be changed.

There is one final refinement to the messaging from our function. We might not want to always print out the messages so we introduce a verbose= option which for now we choose to be TRUE by default. Messages are only printed if verbose=TRUE.

```
perf <- function(mdesc, dstype, verbose=TRUE)
{
  # Provide informative introduction.

  if (verbose)
    "Performance Evaluation\n" %s+%
      "======================\n\n" %s+%
      "Model:    " %s+% mdesc %s+% "\n" %s+%
      "Dataset: " %s+% dstype %s+% "\n" %>%
      cat()
}
perf(mdesc, dstype, verbose=FALSE)
```

9.2.2 Calculating Predictions

The next step is to calculate the predictions. We can review the code used to do this and we will note that each model behaves differently in generating predictions from new datasets. For now we will decide to retain the prediction components of the evaluation outside of our function and will pass the predictions themselves on to the function. We add two more arguments to the function to accept the predicted probabilities and the predicted classes.

```
perf <- function(mdesc, dstype, prob, class,
                 verbose=TRUE)
{
  # Provide informative introduction.

  if (verbose)
    "Performance Evaluation\n" %s+%
      "======================\n\n" %s+%
      "Model:            " %s+% mdesc %s+% "\n" %s+%
      "Dataset:          " %s+% dstype %s+% " dataset with " %s+%
      comma(length(prob)) %s+% " observations.\n" %>%
      cat()
}

model %>%
  predict(newdata=ds[test, vars], type="prob") %>%
  .[,2] %>%
  set_names(NULL) %>%
  round(2) ->
te_prob

model %>%
  predict(newdata=ds[test, vars], type="class") %>%
  set_names(NULL) ->
te_class

perf(mdesc, dstype, te_prob, te_class)

## Performance Evaluation
## ======================
##
## Model:          random forest
## Dataset:        test dataset with 20,217 observations.
```

9.2.3 Accuracy

We now add the first of our evaluations and include some output to report the results. We also expect some data to be returned from the function which is achieved using `base::return()`.

The first evaluation is a simple accuracy calculation. We will compare the predicted class to the actual class (target) and so we now need to also pass the target through to the function. The new argument can be placed in any order but often we choose the order carefully to make some logical sense.

```
perf <- function(mdesc, dstype, target, prob, class,
                 verbose=TRUE)
{
  # Provide informative introduction.

  if (verbose)
    "Performance Evaluation\n" %s+%
      "======================\n\n" %s+%
      "Model:          " %s+% mdesc %s+% "\n" %s+%
      "Dataset:        " %s+% dstype %s+% " dataset with " %s+%
      comma(length(prob)) %s+% " observations.\n" %>%
      cat("\n")

  # Calculate accuracy and error rates.

  sum(class == target, na.rm=TRUE) %>%
    divide_by(class %>% is.na() %>% not() %>% sum()) ->
  acc

  sum(class != target, na.rm=TRUE) %>%
    divide_by(class %>% is.na() %>% not() %>% sum()) ->
  err

  if (verbose)
    "Overall accuracy:  " %s+% percent(acc) %s+% "\n" %s+%
    "Overall error:     " %s+% percent(err) %s+% "\n" %>%
    cat("\n")

  # Return a list of the evaluations.

  return(list(acc=acc, err=err))
}
```

```
perf(mdesc, dstype, te_target, te_prob, te_class)

## Performance Evaluation
## ======================
##
## Model:          random forest
## Dataset:        test dataset with 20,217 observations.
##
## Overall accuracy:  86%
## Overall error:     14%
##
## $acc
## [1] 0.8599512
##
## $err
## [1] 0.1400488
```

The actual returned value is also printed after the function call. If we prefer the results to be returned but not to be printed, then we can use `base::invisible()` in place of `base::return()` which we will do going forward.

9.2.4 Error Matrix

The error matrices come next. The first reports the observation counts whilst the second is expressed as percentages.

```
perf <- function(mdesc, dstype, target, prob, class,
                 verbose=TRUE)
{
  # Provide informative introduction.

  if (verbose)
    "Performance Evaluation\n" %s+%
      "======================\n\n" %s+%
      "Model:            " %s+% mdesc %s+% "\n" %s+%
      "Dataset:          " %s+% dstype %s+% " dataset with " %s+%
      comma(length(prob)) %s+% " observations.\n" %>%
      cat("\n")

  # Calculate accuracy and error rates.

  sum(class == target, na.rm=TRUE) %>%
```

```
    divide_by(class %>% is.na() %>% not() %>% sum()) ->
  acc

  sum(class != target, na.rm=TRUE) %>%
    divide_by(class %>% is.na() %>% not() %>% sum()) ->
  err

  if (verbose)
    "Overall accuracy:  " %s+% percent(acc) %s+% "\n" %s+%
    "Overall error:     " %s+% percent(err) %s+% "\n" %>%
    cat("\n")

  # Generate error matricies.

  matrix <- errorMatrix(target, class)

  if (verbose)
  {
    cat("Error Matrices:\n\n")
    errorMatrix(target, class, count=TRUE) %>% print()
    cat("\n")
    matrix %>% print()
  }

  # Return a list of the evaluations.

  invisible(list(acc=acc,
                 err=err,
                 matrix=matrix))
}

perf(mdesc, dstype, te_target, te_prob, te_class)

## Performance Evaluation
## ======================
##
## Model:          random forest
## Dataset:        test dataset with 20,217 observations.
##
## Overall accuracy:  86%
## Overall error:     14%
##
## Error Matrices:
##
##       Predicted
```

```
## Actual   no  yes Error
##     no 6028  311   4.9
##     yes  836 1015  45.2
##
##        Predicted
## Actual   no  yes Error
##     no 73.6  3.8   4.9
##     yes 10.2 12.4  45.2
```

9.2.5 Performance

We now add in the remaining performance measures to complete our function.

```
perf <- function(mdesc, dstype, target, prob, class,
                 verbose=TRUE)
{
  # Provide informative introduction.

  if (verbose)
    "Performance Evaluation\n" %s+%
      "======================\n\n" %s+%
      "Model:         " %s+% mdesc %s+% "\n" %s+%
      "Dataset:       " %s+% dstype %s+% " dataset with " %s+%
      comma(length(prob)) %s+% " observations.\n" %>%
      cat("\n")

  # Calculate accuracy and error rates.

  sum(class == target, na.rm=TRUE) %>%
    divide_by(class %>% is.na() %>% not() %>% sum()) ->
  acc

  sum(class != target, na.rm=TRUE) %>%
    divide_by(class %>% is.na() %>% not() %>% sum()) ->
  err

  if (verbose)
    "Overall accuracy:  " %s+% percent(acc) %s+% "\n" %s+%
    "Overall error:     " %s+% percent(err) %s+% "\n" %>%
    cat("\n")

  # Generate error matricies.
```

```
matrix <- errorMatrix(target, class)

if (verbose)
{
  cat("Error Matrices:\n\n")
  errorMatrix(target, class, count=TRUE) %>% print()
  cat("\n")
  matrix %>% print()
  cat("\n")
}

# Calculate recall, precision and F-score.

rec <- (matrix[2,2]/(matrix[2,2] + matrix[2,1]))
pre <- (matrix[2,2]/(matrix[2,2] + matrix[1,2]))
fsc <- ((2 * pre * rec)/(rec + pre))

if (verbose)
  "Recall:    " %s+% percent(rec) %s+% "\n"  %s+%
  "Precision: " %s+% percent(pre) %s+% "\n"  %s+%
  "F-Score:   " %s+% round(fsc, 3) %s+% "\n" %>%
  cat("\n")

# Calculate AUC for the ROC curve.

prob %>%
  prediction(target) %>%
  performance("auc") %>%
  attr("y.values") %>%
  .[[1]] ->
auc

if (verbose)
  "Percentage area under the ROC curve AUC: " %s+%
    percent(auc) %>%
  cat("\n")

prob %>%
  prediction(target) %>%
  performance("tpr", "fpr") ->
rates

# Return a list of the evaluations.
```

```
  invisible(list(acc=acc,
                 err=err,
                 matrix=matrix,
                 rec=rec,
                 pre=pre,
                 auc=auc,
                 rates=rates))
}

m_rf_perf <- perf(mdesc, dstype, te_target, te_prob, te_class)

## Performance Evaluation
## ======================
##
## Model:         random forest
## Dataset:       test dataset with 20,217 observations.
##
## Overall accuracy:  86%
## Overall error:     14%
##
## Error Matrices:
##
##       Predicted
## Actual   no  yes Error
##    no  6028  311   4.9
##    yes  836 1015  45.2
##
##       Predicted
## Actual   no  yes Error
##    no  73.6  3.8   4.9
##    yes 10.2 12.4  45.2
##
## Recall:    54.9%
## Precision: 76.5%
## F-Score:   0.639
##
## Percentage area under the ROC curve AUC: 89.8%
```

9.3 Function for ROC Curves

Another smaller piece of code we found ourselves repeating throughout the previous chapters was the code for the ROC curve. Here we package that code into a function having four arguments. The first argument is the so-called rates generated through the evaluation. This is effectively the x and y points for the ROC curve. We also pass in the actual calculated value of the AUC and then the two descriptive strings for the model and the dataset.

```
aucplot <- function(rates, auc, mdesc, dstype)
{
  data.frame(tpr=attr(rates, "y.values")[[1]],
             fpr=attr(rates, "x.values")[[1]]) %>%
    ggplot(aes(fpr, tpr)) +
    geom_line() +
    labs(title="ROC - " %s+% mtype %s+% " - Training Dataset",
         subtitle=paste("AUC =", percent(auc)),
         x="False Positive Rate (1-Specificity)",
         y="True Positive Rate (Sensitivity)")
}

aucplot(m_rf_perf$rates, m_rf_perf$auc, mdesc, dstype)
```

The resulting plot can be seen in Figure 9.1. Notice that we do not have an explicit base::return() for the function. R automatically returns the result of the last function within the function as its return value. In this case we return the plot object. This is useful in the case of ggplot objects since we can further add to the plot outside of the function. Here for example we add a caption to the plot.

```
aucplot(m_rf_perf$rates, m_rf_perf$auc, mdesc, dstype) +
  labs(caption="Generated 2017-07-20")
```

Functions are a powerful concept and we will find many opportunities for writing functions.

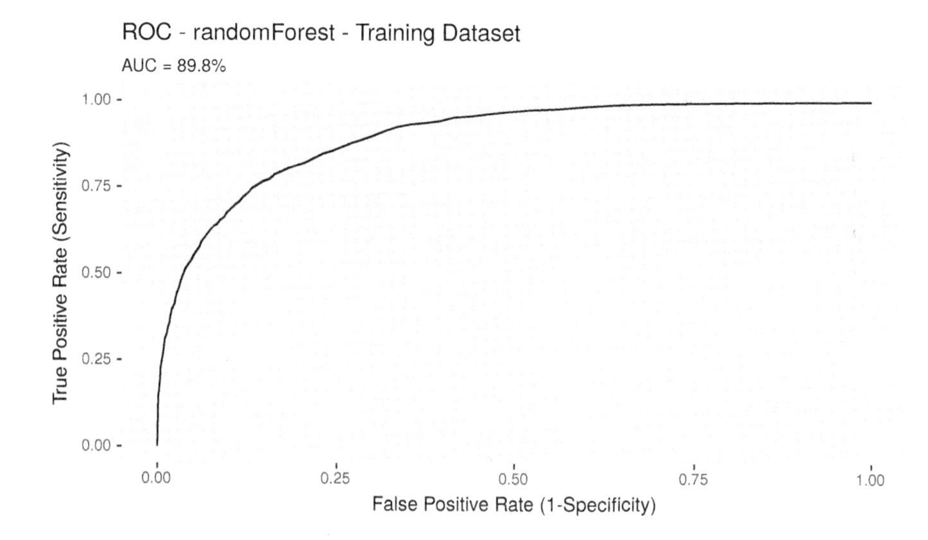

Figure 9.1: *ROC curve plotted using our own aucplot().*

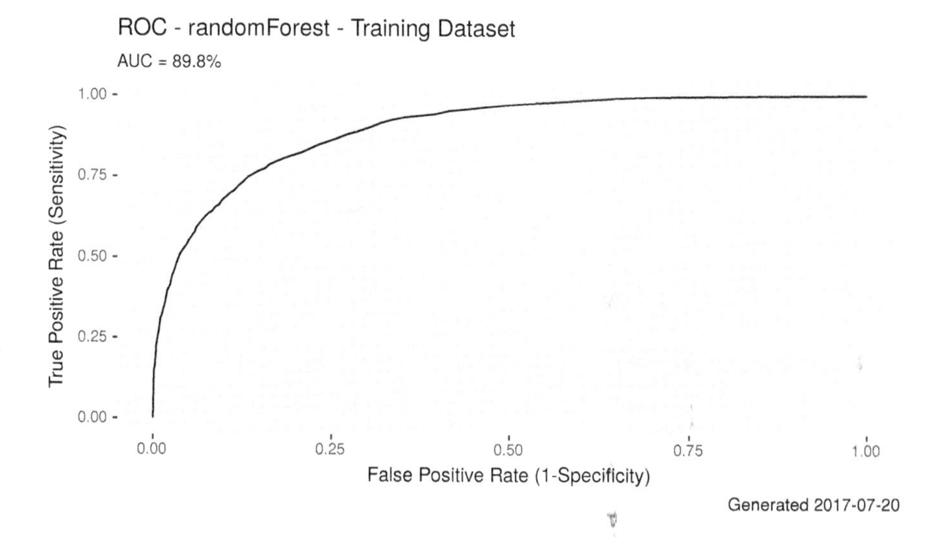

Figure 9.2: *ROC curve with a caption.*

9.4 Exercises

Exercise 9.1 Variable Weights

Write a function to take a dataset and return a vector of probabilities related to the correlation of a variable with the target, one for each input variable in the dataset. Higher correlation will lead to a higher probability.

R's correlation functions can be used to calculate the correlation between each column (variable) of the data frame and the values of the target variable. Below are some hints.

```
n1 <- ds[["temp_3pm"]]
c1 <- ds[["wind_gust_dir"]]
t1 <- ds[[target]]

cor(as.numeric(n1), as.numeric(t1), use="pairwise.complete.obs")

## [1] -0.1960651

cor(as.numeric(c1), as.numeric(t1), use="pairwise.complete.obs")

## [1] 0.06100805
```

The template for the function is

```
varWeights <- function(ds, target)
{ ... }
```

The solution will produce the following output where the correlations are mapped into probabilities between 0 and 1.

```
varWeights(ds, target)

##            date        location        min_temp
##    0.0021740132    0.0010802198    0.0170236723
##        max_temp        rainfall     evaporation
##    0.0356162094    0.0515446458    0.0265293351
##        sunshine   wind_gust_dir wind_gust_speed
##    0.0984482796    0.0132820139    0.0502786994
....
```

10

Literate Data Science

A data scientist's role is to tell the stories supported by the data we are analysing. The narrative that we tell is one of our key deliverables and as such we need our narrative to be well supported by the data. In telling the narrative we would like the analysis to be transparent, repeatable, and reproducible. We would like to capture and share our activities to ensure the quality of our work and for peer review. We will also find ourselves repeating our work on other datasets in other scenarios and with other organisations. Documenting what we do helps when we come back to the code at a later time. Others will also want to reproduce our work and we should do all we can to facilitate that process. In short, we need to clearly communicate what we do so that we and others can understand and can continue the journey.

A general rule of thumb tells us that we should spend about a quarter of our time capturing what we have done—documenting our projects. Even more important is to capture this as we are doing the work rather than having the chore to write up our work later on. This does present an overhead and risks interrupting the flow of our work but the investment pays off longer term. Tools are available that support the capture of our work with minimal interruption to our work flow.

To support the narrative and to encourage our efforts to be transparent, repeatable, and reproducible, we deploy the concept of **literate programming** (Knuth, 1984). The concept is to intermix our narrative with the underlying analyses of the data (our code) within the one document. By introducing the concept here we aim to provide a solid foundation for the data scientist. We won't always have the time or the patience to deliver a carefully crafted narrative telling the story derived from the data but we should strive to do so.

We will use knitr (Xie, 2016) to support literate data science. The package combines the document typesetting power of the free and open source LaTeX software with the statistical power of R. Literate data science is also well supported by RStudio which is able to process the source document into a beautifully formatted PDF. This book is itself produced using knitr and LaTeX.

LaTeX is not the only option and indeed today many beginners learn literate programming using markdown or R markdown as supported by RStudio. From an R markdown document we can create documents in a variety of other formats, including LaTeX and also interactive Jupyter Notebooks.

Packages used in this chapter include Hmisc (Harrell, 2017), diagram (Soetaert, 2014), dplyr (Wickham *et al.*, 2017a), ggplot2 (Wickham and Chang, 2016), magrittr (Bache and Wickham, 2014), rattle (Williams, 2017) and xtable (Dahl, 2016).

```
# Load required packages from the local library into R session.

library(Hmisc)    # Escape special LaTeX charaters.
library(diagram)  # Produce a flowchart.
library(dplyr)    # Data wrangling: tbl_df().
library(ggplot2)  # Visualise data.
library(magrittr) # Pipelines for data processing: %>% %T>% %<>%.
library(rattle)   # The weatherAUS dataset.
library(xtable)   # Format R data frames as LaTeX tables.
```

In addition to these packages we also need to install the LaTeX software. LaTeX is a typesetting markup language which combined with knitr allows us to intermix R code with our narrative and to program certain parts of the narrative using R. Like R, LaTeX is free and open source software and instructions for installing are available from the LaTeX Project.*

*https://latex-project.org

10.1 Basic LaTeX Template

LaTeX is also a language for programming just like R. With LaTeX we program over words rather than program over data. It is what we call a markup language and has its own collection of commands that tell the LaTeX software what to do. Within RStudio when we create a new knitr document a skeleton of LaTeX commands will be automatically inserted into the document for us. From this initial script we can readily produce well-formatted reports and presentations.

Once we have a new text document with a filename extension of .Rnw we will enter the LaTeX commands intermixed with the text of our narrative intermixed with the R commands performing the data wrangling and the analyses.

In creating a new knitr document RStudio will insert the following template as the minimal LaTeX markup.[*] We can see the RStudio context in Figure 10.1. The file contains the following minimal LaTeX code.

```
\documentclass{article}

\begin{document}

\end{document}
```

Into this file we will begin entering the LaTeX commands and the text of our narrative and the R code. We enter text to describe our project between the begin and end of the document and then choose the `Compile PDF` button on the toolbar to generate a PDF document.

LaTeX provides an extensive collection of commands including the basics to highlight text (for example, `\bold{...}` or `\italic{...}`) and to create a list of items:

---

[*] In RStudio we create a new R Sweave document under the File menu. We need to inform RStudio to use knitr which is a significantly more advanced version of Sweave. Under the Tools menu choose Global Options and then Sweave to set the Weave option to be `knitr`.

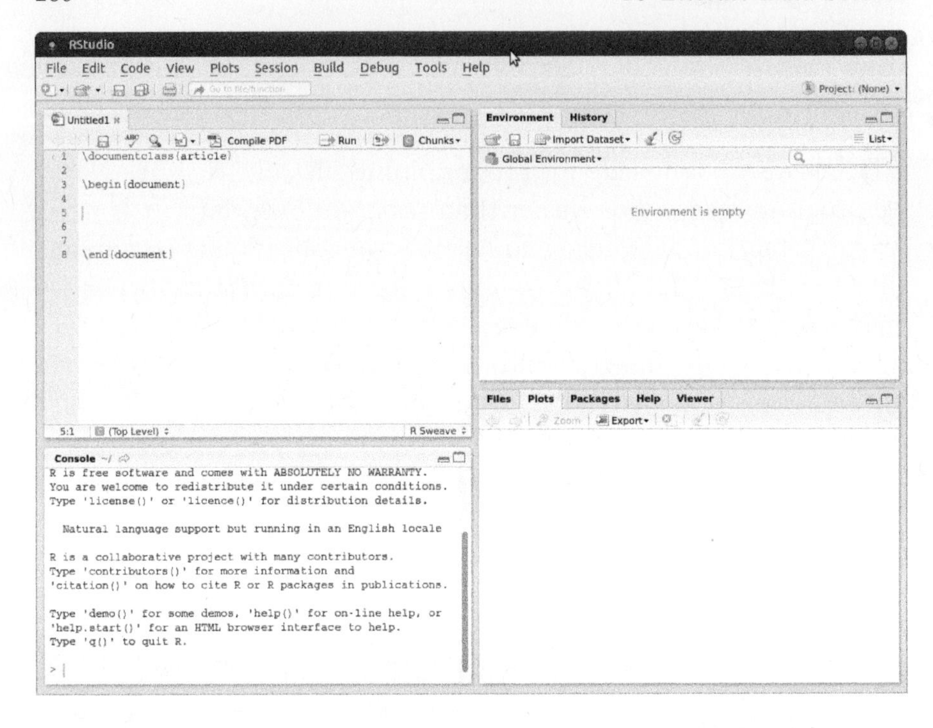

Figure 10.1: *Creating a new R Sweave document in RStudio.*

```
\begin{itemize}
\item Text for the \bold{first} item.
\item Text for the \italic{second} item.
\item ...
\end{itemize}
```

We will see many examples as we proceed through this chapter and there are many guides to LaTeX available on the Internet. Our aim is to quickly be comfortable with using LaTeX and not to let it get in our way.

10.2 A Template for our Narrative

A basic template can be the starting point for any new project. We provide a template here as a complete LaTeX document. We can

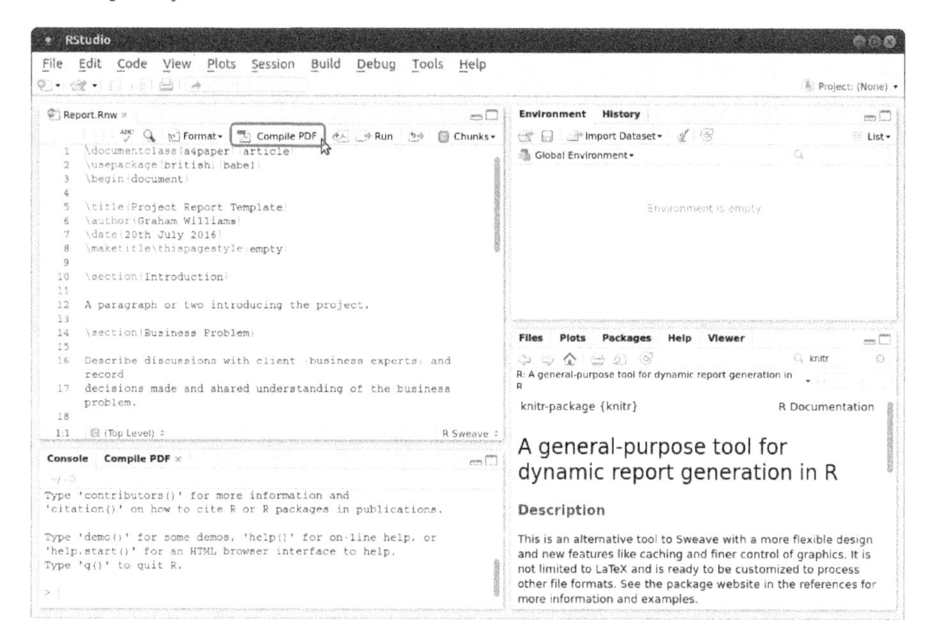

Figure 10.2: *Within the* RStudio *edit pane we are editing the LaTeX document. A click of the* `Compile PDF` *button will run* ***knitr*** *over the document and then run the LaTeX software to produce a PDF document.*

simply copy it into RStudio and begin creating a narrative.* Be sure to save the file using the .Rnw filename extension to identify it as an Sweave/knitr document. We can then generate a typeset PDF document simply by clicking the `Compile PDF` button in RStudio.

In Figure 10.2 we see the interaction with RStudio showing the template within the editor in the top left window pane. When we have processed the document (compiled to PDF) we will see the formatted document as in Figure 10.3.

The template is simply:

```
\documentclass[a4paper]{article}
\usepackage[british]{babel}
\begin{document}

\title{Project Report Template}
```

*The template is available from `https://essentials.togaware.com`.

Project Report Template

Graham Williams

2nd July 2017

1 Introduction

A paragraph or two introducing the project.

2 Business Problem

Describe discussions with client (business experts) and record decisions made and shared understanding of the business problem.

3 Data Sources

Identify the data sources and discuss access with the data owners. Document data sources, integrity, providence, and dates.

4 Data Preparation

Load the data into R and perform various operations on the data to shape it for analysis.

5 Data Exploration

We should always understand our data by exploring it in various ways. Include data summaries and various plots that give insights.

6 Data Analytics

Include all models built and parameters tried. Include R code and model evaluations.

7 Deployment

Choose the model to deploy and export it, perhaps as PMML.

Figure 10.3: *The PDF generated from the template* **knitr** *report.*

```
\author{Graham Williams}
\date{2nd July 2017}
\maketitle\thispagestyle{empty}

\section{Introduction}

A paragraph or two introducing the project.

\section{Business Problem}

Describe discussions with client (business experts)
and record decisions made and shared understanding of
the business problem.
```

```
\section{Data Sources}

Identify the data sources and discuss access with the
data owners. Document data sources, integrity,
providence, and dates.

\section{Data Preparation}

Load the data into R and perform various operations
on the data to shape it for analysis.

\section{Data Exploration}

We should always understand our data by exploring it
in various ways. Include data summaries and various
plots that give insights.

\section{Data Analytics}

Include all models built and parameters
tried. Include R code and model evaluations.

\section{Deployment}

Choose the model to deploy and export it, perhaps as
PMML.

\end{document}
```

10.3 Including R Commands

We can include R commands within the knitr document and have the commands automatically run when we compile the PDF. The output from the R commands will be displayed together with the R commands themselves.

To include R commands we surround the code with special markers. The code blocks containing the commands begin with double less than symbols (or angle brackets <<) starting in column one and end with double greater than symbols (>>) followed immediately by an equals (=). The code block is then terminated by a line containing a single “at” symbol (@) starting in column one.

```
<<>>=
... R code ...
@
```

Between the angle brackets we place instructions to tell knitr what to do with the R commands. We can tell it to simply echo the commands, but not to evaluate them, or to evaluate the commands without echoing them, and so on. Whilst it is optional, we should provide a label for each block of R code. This is the first element between the angle brackets. A simple example of the beginning of a typical code block is:

```
<<my_label, eval=TRUE, echo=FALSE>>=
```

The label here is `my_label` and we ask knitr to evaluate the R commands and thus to also show the output of those commands (this is the default). We do not echo the R commands so that the actual commands themselves will not appear in the resulting document (the default is to echo the commands). Whilst we develop our narrative we will include all of the code chunks and the output into the generated PDF report but once we are ready to produce our final report we would turn the echoing of the R code off, globally.

Our first example of running actual R code will generate some random uniform data using `stats::runif()` and then view the `utils::head()` of the data and calculate the `base::mean()`. The following code block shows how this will look in the source .Rnw file.

```
<<example_random_mean>>=
# Always include a short comment to support the code.
x <- runif(1000) * 1000
head(x)
mean(x)
@
```

Below is what it looks like after it is processed by knitr and then LaTeX (as happens when we click the `Compile PDF` button in RStudio).

```
# Always include a short comment to support the code.

x <- runif(1000) * 1000
head(x)

## [1] 914.8060 937.0754 286.1395 830.4476 641.7455 519.0959

mean(x)

## [1] 488.2555
```

Notice that the syntax is colour highlighted and the output is included as comments introduced in R with the `##`. If we were to evaluate these commands ourselves in R, the output would not include the `##`.

10.4 Inline R Code

Often we find ourselves wanting to refer to the results of an R command within the text we are writing rather than as a separate code chunk. This is easily done using the `\Sexpr{}` command in LaTeX.

To include today's date within the narrative we can type the command sequence exactly as `\Sexpr{Sys.Date()}`. This will be replaced with "2017-07-20". Any R function can be called in this way. To format the date, for example, we can use R's `base::format()` function to specify how the date is displayed, as in \Sexpr{Sys.Date() %>% format(format="%A, %e %B %Y")} to produce Sunday, 2 July 2017.

We typically intermix a narrative of our dataset with output from R to support and illustrate the discussion. In the following sentence we do this showing first the output from the R command and then the actual R command we included in the source document. For example, the **weather** dataset from rattle (Williams, 2017) has 366 (i.e., \Sexpr{nrow(weather)}) observations including observations of the following 4 of the 24 (i.e., \Sexpr{ncol(weather)}) available variables: MinTemp, MaxTemp, Rainfall, Evaporation (i.e., \Sexpr{names(weather)[3:6] %>% paste(collapse=", ")}).

LaTeX treats some characters specially and we need to be careful to escape such characters. For example, the underscore "_" is used to introduce a subscript in LaTeX. It needs to be escaped if we really want an underscore to be included in the compiled PDF document. If not, LaTeX will likely complain. As an example, we might list one of the variable names from the **weather** dataset with an underscore in its name: RISK_MM (\Sexpr{names(weather)[23]}). We will see an error like:

```
KnitR.tex:230: Missing $ inserted.
KnitR.tex:230: leading text: ...an underscore in its name: RISK_
KnitR.tex:232: Missing $ inserted.
```

Hmisc::latexTranslate() assists here. It was used above to print the variable name. There are other support functions in Hmisc that are useful when working with knitr and LaTeX—see library(help=Hmisc) for further information.

10.5 Formatting Tables Using Kable

Including a typeset table based on a dataset can be accomplished using knitr::kable(). Here we will use the larger **weatherAUS** dataset from rattle setting it up as a dplyr::tbl_df(). We will then choose specific columns and a random selection of rows to include in the table. The source text we include in our .Rnw file is listed in the following code block.

```
<<example_kable, echo=TRUE, results="asis">>=
# Set the seed so that results are repeatable.
set.seed(42)
# Load package from local library into the R session.
library(rattle)
# Record metadata for a sample of the dataset.
nobs <- nrow(weatherAUS)
obs  <- sample(nobs, 5)
vars <- 2:6
ds   <- weatherAUS[obs, vars]
# Generate the appropriate LaTeX code to display the data.
kable(ds)
@
```

The result (also showing the R code since we specified echo=TRUE) is then:

```
# Set the seed so that results are repeatable.

set.seed(42)

# Load package from local library into the R session.

library(rattle)

# Record metadata for a sample of the dataset.

nobs <- nrow(weatherAUS)
obs  <- sample(nobs, 5)
vars <- 2:5
ds   <- weatherAUS[obs, vars]

# Generate the appropriate LaTeX code to display the
 data.

kable(ds)
```

	Location	MinTemp	MaxTemp	Rainfall
126525	Hobart	9.3	17.9	0.0
129604	AliceSprings	23.0	37.3	0.2
39575	Williamtown	10.5	18.2	1.0
114855	Perth	5.0	20.1	0.0
88756	Townsville	24.4	31.4	23.6

Since we are working with a random sample and we would like the sampling to be repeatable we have used `base::set.seed()` to initialise the random number generator to a fixed value.

Formatting Options

Formatting options are available for fine tuning how the table is to be presented. For example we can remove the row names (the row numbers in the above table) easily with `row.names=FALSE`.

```
# Display a table without row names.

ds %>% kable(row.names=FALSE)
```

Location	MinTemp	MaxTemp	Rainfall
Hobart	9.3	17.9	0.0
AliceSprings	23.0	37.3	0.2
Williamtown	10.5	18.2	1.0
Perth	5.0	20.1	0.0
Townsville	24.4	31.4	23.6

We can also limit the number of digits displayed to avoid an impression of a high level of accuracy or to simplify presentation using `digits=`. By doing so the numeric values are rounded to the requested number of decimal points.

```
# Display a table removing digits from numbers.

ds %>% kable(row.names=FALSE, digits=0)
```

Location	MinTemp	MaxTemp	Rainfall
Hobart	9	18	0
AliceSprings	23	37	0
Williamtown	10	18	1
Perth	5	20	0
Townsville	24	31	24

Improvements Using BookTabs

The `booktabs` package for LaTeX provides additional functionality that we can make use of with `knitr::kable()`. To use this be sure to include the following in the preamble (before the `\begin{document}` of your .Rnw file:

```
# Load package from local library into the R session.

\usepackage{booktabs}
```

We can then set `booktabs=TRUE` to remove the clutter of the extra lines.

```
# Use booktabs option to improve presentation of table.

ds %>% kable(row.names=FALSE, digits=0, booktabs=TRUE)
```

Location	MinTemp	MaxTemp	Rainfall
Hobart	9	18	0
AliceSprings	23	37	0
Williamtown	10	18	1
Perth	5	20	0
Townsville	24	31	24

In the following example we notice that with more rows `booktabs=TRUE` will add a small gap every 5 rows.

```
# Display a tale with more observations.

weatherAUS[sample(nobs, 20), vars] %>%
  kable(row.names=FALSE, digits=0, booktabs=TRUE)
```

Location	MinTemp	MaxTemp	Rainfall
Portland	8	16	0
Woomera	19	36	0
NorahHead	18	18	1
Townsville	27	34	0
MountGambier	10	16	6
MelbourneAirport	10	21	0
Nuriootpa	6	14	0
Launceston	9	16	1
WaggaWagga	0	8	0
MelbourneAirport	7	28	0
AliceSprings	11	24	0
Darwin	22	34	0
Newcastle	10	22	1
Melbourne	8	19	NA
Dartmoor	12	20	0
Hobart	3	9	2
NorahHead	13	23	0
Katherine	26	36	0
AliceSprings	14	32	0
CoffsHarbour	17	30	1

10.6 Formatting Tables Using XTable

Whilst `knitr::kable()` provides basic functionality much more extensive control over the formatting of tables is provided by xtable (Dahl, 2016). By default the table produced is called a floating table so that it automatically floats within the document to an appropriate location.

As a floating table we will add a `caption=` and a table reference `label=` to the table so that they do not get lost. We can then refer to the table within the text and have the tables appear somewhere convenient automatically. The code block below, for example, produces Table 10.1 on page 271. The table and page numbers are automatically assigned to the table. Within LaTeX we can access the table number using `\ref{egtbl}` (10.1) and the page number using `\pageref{egtbl}` (271).

```
# Load package from local library into the R session.

library(xtable)

# Generate a floating table with a caption.

ds %>%
  xtable(caption="Example xtable.", label="egtbl") %>%
  print(caption.placement="top")
```

Table 10.1: *Example xtable.*

	Location	MinTemp	MaxTemp	Rainfall
126525	Hobart	9.30	17.90	0.00
129604	AliceSprings	23.00	37.30	0.20
39575	Williamtown	10.50	18.20	1.00
114855	Perth	5.00	20.10	0.00
88756	Townsville	24.40	31.40	23.60

Also note that by default missing values (`NA`) are not printed nor are the extra lines that are printed by default when using `knitr::kable()`.

There are many formatting options available for fine tuning how the table is to be presented and we cover some of these in the following pages. We also note that some options are provisioned by `xtable::xtable()` whilst others are available through `xtable::print.xtable()`. An example option is `include.rownames=` which is an option available with `xtable::print.xtable()`. The result is seen in Table 10.2.

```
# Display a table without row names.

ds %>%
```

```
  xtable(caption = "Remove row numbers.",
         label   = "tblnonums") %>%
  print(caption.placement = "top",
        include.rownames  = FALSE)
```

Table 10.2: *Remove row numbers.*

Location	MinTemp	MaxTemp	Rainfall
Hobart	9.30	17.90	0.00
AliceSprings	23.00	37.30	0.20
Williamtown	10.50	18.20	1.00
Perth	5.00	20.10	0.00
Townsville	24.40	31.40	23.60

Formatting Numbers with XTable

As with `knitr::kable()` we can limit the number of digits displayed to avoid giving an impression of a high level of accuracy or to simplify the presentation. In Table 10.3 we have removed all decimal points.

```
# Display a table removing digits from numbers.

ds %>%
  xtable(digits  = 0,
         caption = "Decimal points.",
         label   = "tbldp0") %>%
  print(caption.placement = "top",
        include.rownames  = FALSE)
```

When we have large numbers being displayed it is imperative that we include commas to separate the thousands. Many mistakes are made misreading numbers that include many digits when commas are not included.

Table 10.3: *Decimal points.*

Location	MinTemp	MaxTemp	Rainfall
Hobart	9	18	0
AliceSprings	23	37	0
Williamtown	10	18	1
Perth	5	20	0
Townsville	24	31	24

```
# Copy dataset so as to change the data.

dst     <- ds

# Randomly create large numbers.

dst[-1] <- sample(10000:99999, nrow(dst)) * dst[-1]

# Illustrate default table display of large numbers.

dst %>%
  xtable(digits  = 0,
         caption = "Large numbers.",
         label   = "tbllrg") %>%
  print(caption.placement = "top",
        include.rownames  = FALSE)
```

Table 10.4: *Large numbers.*

Location	MinTemp	MaxTemp	Rainfall
Hobart	523395	1007394	0
AliceSprings	1037691	1682864	9023
Williamtown	960897	1665555	91514
Perth	251125	1009523	0
Townsville	2079783	2676442	2011593

Consider the result in Table 10.4. It is difficult to distinguish between the thousands and millions. We often find ourselves having to carefully count the digits to check whether 1007394 really is 1,007,394. To avoid this cognitive load on the reader, we should always use a comma to separate the thousands and millions. This simple principle makes it much easier for the reader to appreciate the scale and to avoid misreading data, yet it is so often overlooked. We can see the result in Table 10.5.

```
# Format large numbers using commas as appropriate.

dst %>%
  xtable(digits  = 0,
         caption = "Large numbers formatted.",
         label   = "tbllrgf") %>%
  print(caption.placement = "top",
        include.rownames  = FALSE,
        format.args       = list(big.mark=","))
```

Table 10.5: *Large numbers formatted.*

Location	MinTemp	MaxTemp	Rainfall
Hobart	523,395	1,007,394	0
AliceSprings	1,037,691	1,682,864	9,023
Williamtown	960,897	1,665,555	91,514
Perth	251,125	1,009,523	0
Townsville	2,079,783	2,676,442	2,011,593

Sophisticated Captions

Captions can be formatted with a little knowledge of LaTeX. For Table 10.6 we illustrate generating a string in R that is passed through to the `caption=`. We use `base::paste()` and `base::Sys.time()` and include some special symbols known to LaTeX as well as an occasional bold and italic font. Notice that because the caption is quite long we do not want the whole caption included in the list of tables in the contents pages. The second argument to `caption=` is the short title to use for the tables of contents.

```
# Create a long caption as a single srting.

cpt <- paste("Here include in the \\textbf{caption}",
             "a sample of \\LaTeX{} symbols and",
             "formats that can be included in the",
             "string, and note that the caption",
            "string can be the result of R commands,",
             "using \\texttt{paste()} in this",
             "instance. Some sample symbols include:",
             "$\\alpha$ $\\rightarrow$ $\\wp$.",
             "We also get a timestamp from R:",
             paste0(Sys.time(), "."))

# Add the caption to the table.

dst %>%
  xtable(digits  = 0,
         caption = c(cpt, "Extended caption."),
         label   = "tblcap") %>%
  print(caption.placement = "top",
        include.rownames  = FALSE)
```

Table 10.6: *Here we include in the* ***caption*** *a sample of LaTeX symbols and formats that can be included in the string, and note that the caption string can be the result of R commands, using* `paste()` *in this instance. Some sample symbols include:* $\alpha \rightarrow \wp$. *We also get a timestamp from R: 2017-07-20 15:00:00.*

Location	MinTemp	MaxTemp	Rainfall
Hobart	523395	1007394	0
AliceSprings	1037691	1682864	9023
Williamtown	960897	1665555	91514
Perth	251125	1009523	0
Townsville	2079783	2676442	2011593

10.7 Including Figures

Plots and other graphics can be the difference between a dull report that few will ever read and an exciting report that invites the reader into an interesting narrative. Through knitr we can readily include graphics that can be generated by R. We will illustrate with simple examples.

Sample Figure

To include figures generated by R in our document we simply add plotting commands to the code chunk. Here, for example, is R code to generate a simple density plot of the 3pm temperature in four cities over a year. We use ggplot2 (Wickham and Chang, 2016) to generate the figure.

```
# Load packages from the local library into the R session.

library(rattle)    # For the weatherAUS dataset.
library(ggplot2)   # To generate a density plot.

# Identify cities of interest.

cities <- c("Canberra", "Darwin", "Melbourne", "Sydney")

# Generate the plot.

weatherAUS %>%
  subset(Location %in% cities & ! is.na(Temp3pm)) %>%
  ggplot(aes(x=Temp3pm, colour=Location, fill=Location)) +
  geom_density(alpha=0.55)
```

In the source document (the .Rnw file) the above R code is actually inserted between the chunk begin and end marks within the document itself. Those marks are

```
<<myfigure, eval=FALSE>>=
... R code ...
@
```

Notice the use of `eval=FALSE`, which allows the R code to be included in the text of the final document, as it is above, but will not yet generate the plot to be included in the figure. We leave that for a little later.

The code chunk begins by attaching the requisite packages: rattle (Williams, 2017) to access the **weatherAUS** dataset and ggplot2 (Wickham and Chang, 2016) for the function to generate the actual plot.

The four cities we wish to plot are then identified and we generate a `base::subset()` of the **weatherAUS** dataset containing just those cities. We pass the subset on to `ggplot2::ggplot()` and identify `Temp3pm` for the x-axis, using `location` to colour and fill the plot. We add a layer to the figure containing a density plot with a level of transparency specified as an `alpha=` value. We can see the figure below.

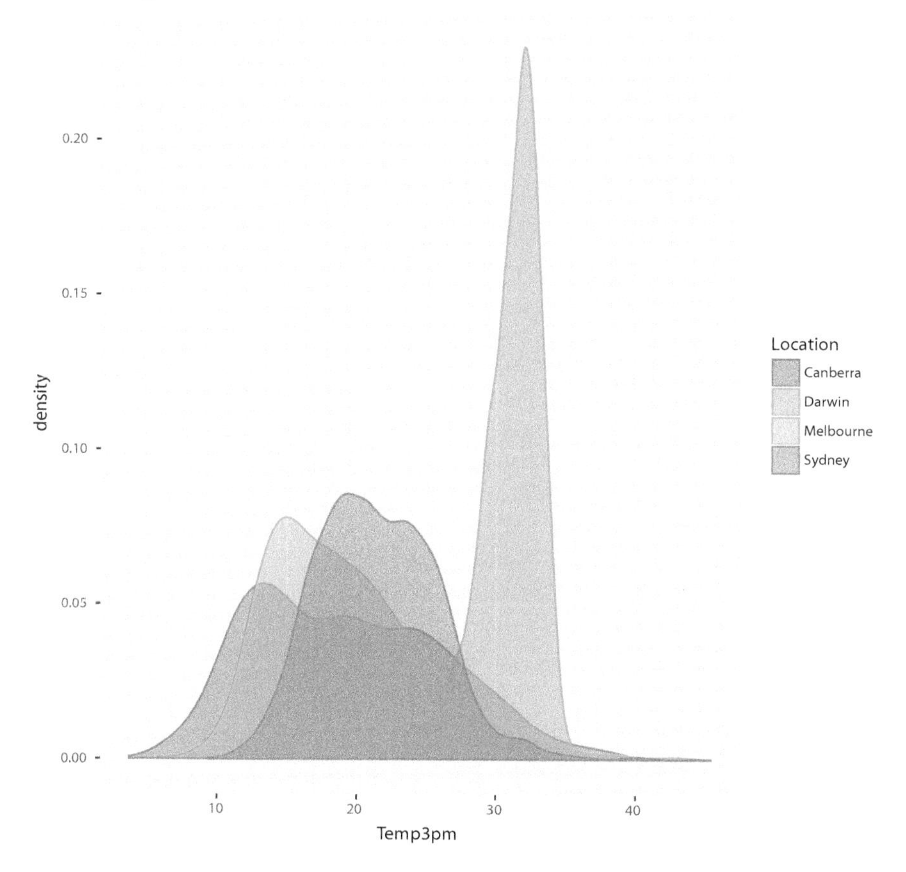

We include the figure in the final document as above simply by removing the eval=FALSE from the previous code chunk. Thus, the R code is evaluated and a plot is generated. We have actually replaced the eval=FALSE with echo=FALSE so as not to print the R code a second time.

We do not actually need to rewrite the R code again in a second chunk, given the code has already been provided in the first chunk on the previous page. We use a feature of knitr where an empty chunk having the same name as a previous chunk is actually a reference to that previous chunk. Thus in our source .Rnw text document we add the following two lines. This is effectively replaced by the R code from the previous block of the same name.

```
<<myfigure, echo=FALSE>>=
@
```

This is exactly what we included at the beginning of this section in the actual source document for this page. Noticing that we have replaced eval=FALSE with echo=FALSE, we cause the original R code to be executed, generating the plot which is included as the previous figure. Using echo=FALSE simply ensures we do not include the R code itself in the final output, this time. That is, the R code is replaced with the figure it generates.

Notice how the figure takes up quite a bit of space on the page.

Adjusting Aspect

We can fine-tune the size of the figure to suit the document and presentation. In this example we have asked R to widen the figure from 7 inches to 14 inches using fig.width=. The code chunk is:

```
<<myfigure, echo=FALSE, fig.width=14>>=
... R code ...
@
```

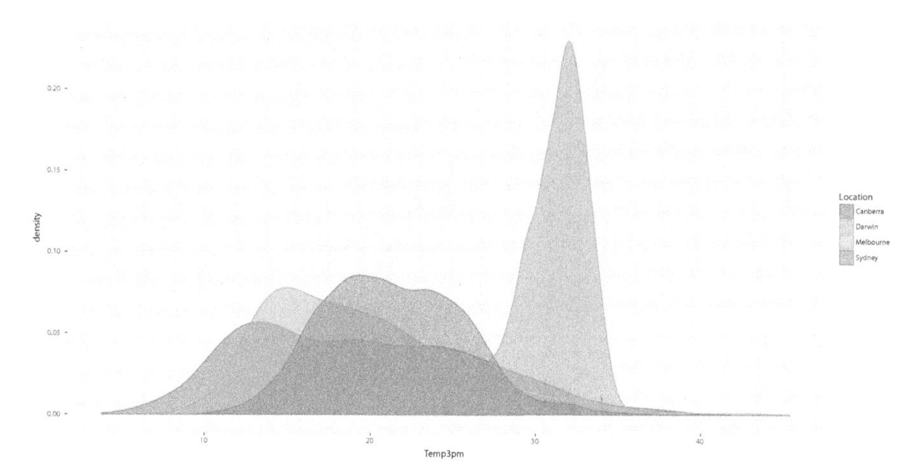

Underneath, knitr is using a PDF device on which the plot is generated, and then saved to file for inclusion in the final document. The PDF device grDevices::**pdf**(), by default, will generate a 7 inch by 7 inch plot (see ?pdf for details). This is the plot dimensions as we saw earlier. By setting fig.width= (and fig.height=) we can change the dimensions. In our example here we have doubled the width, resulting in a more pleasing plot.

Notice that as a consequence of the figure being larger the fonts have remained the same size, resulting in them appearing smaller now when we include the figure in the same area on the printed page.

Choosing Dimensions

Often a bit of trial and error is required to get the dimensions right. Notice though that increasing the fig.width= as we did for the previous plot, and/or increasing the fig.height=, effectively also reduces the font size. Actually, the font size remains constant whilst the figure grows (or shrinks) in size. Sometimes it is better to reduce the fig.width or fig.height to retain a good sized font.

The plot below was generated with the following knitr options.

```
<<myfigure, echo=FALSE, fig.height=3.5>>=
... R code ...
@
```

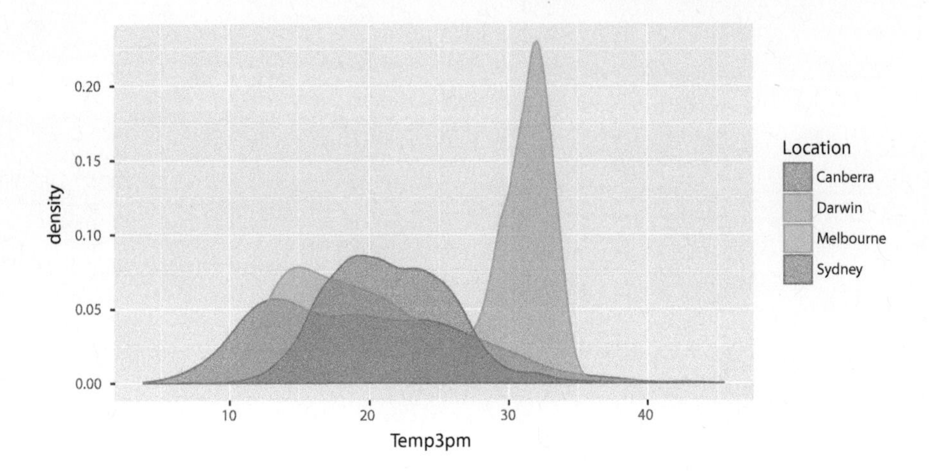

Setting Output Width

We can use the `out.width=` and `out.height=` to adjust how much space a figure takes up in the final document. The below figure is reduced to fill just half the textwidth of the document using:

```
<<myfigure, out.width="0.5\\textwidth", fig.align="center", f...
... R code ...
@
```

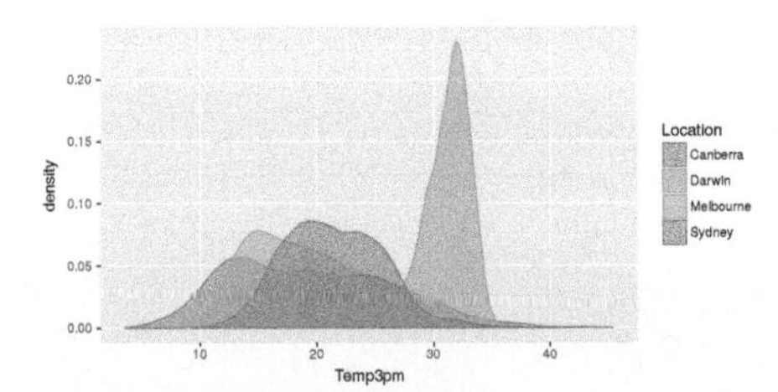

If that is too narrow we can increase it to 90% of the page width with:

```
<<myfigure, out.width="0.9\\textwidth", fig.align="center", f...
... R code ...
@
```

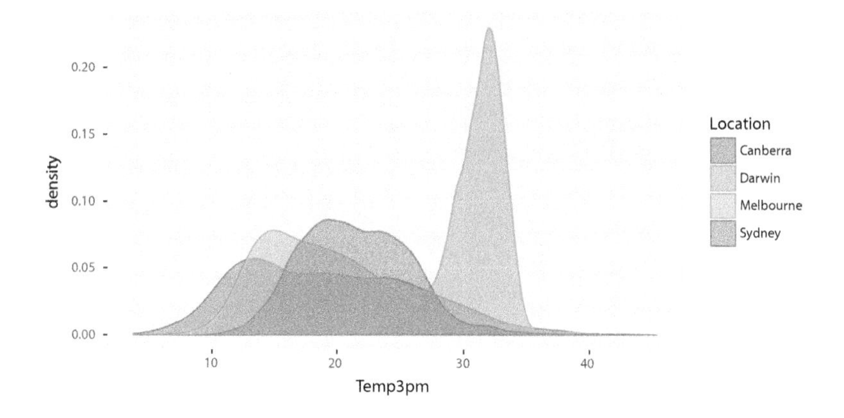

10.8 Add a Caption and Label

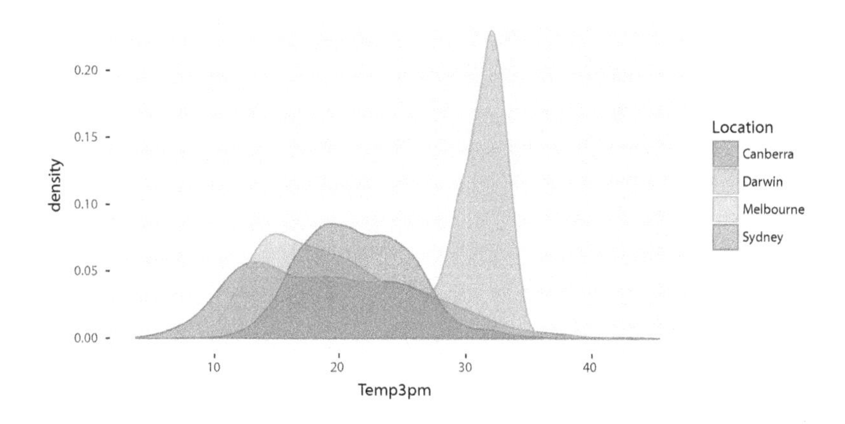

Figure 10.4: *The 3pm temperature for four locations.*

Adding a caption (which automatically also adds a label) is done using `fig.cap=`.

```
<<myfigure, fig.cap="The 3pm temperature for four locations."...
... R code ...
@
```

We have also used `fig.pos="h"` which requests placement of

the figure "here" rather than letting it float. Other options are to place the figure at the top of a page ("t"), or the bottom of a page ("b"). We can leave it empty and the placement is done automatically—that is, the figure floats to an appropriate location.

Once a caption is added, a label is also added to the figure so that it can be referred to in the document. The label is made up of `fig:` followed by the chunk label, which is `myfigure` in this example. So we can refer to the figure using `\ref{fig:myfigure}` and `\pageref{fig:myfigure}`, which allows us to refer to Figure 10.4 on Page 281.

10.9 Knitr Options

We have explored only a small portion of the functionality of knitr and LaTeX. Together there is virtually nothing they cannot do—being programming languages anything that can be done, can be done using these tools.

As a summary below we list the common knitr options. The options are added to the chunk start line or else we can set the options using `opts_chunk$set()`. The arguments to the function can be any number of named options with their values. For example:

```
# Set global defaults for knitr options.

opts_chunk$set(size="footnotesize", message=FALSE, tidy=FALSE)
```

Once this is run, the options remain in force as the default values until they are again changed using `opts_chunk$set()`. They can be overriden per chunk in the usual way.

```
background="#F7F7F7"      # Background colour of the code chunks.
cache.path="cache/"       #
comment=NA                # Suppresses "\verb|## |" in R output.
echo=FALSE                # Do not show R commands.
echo=3:5                  # Only echo lines 3 to 5 of the chunk.
eval=FALSE                # Do not run the R code.
eval=2:4                  # Evaluate lines 2 to 4 of the chunk.
```

```
fig.align="center"          #
fig.cap="Caption..."        #
fig.keep="high"             #
fig.lp="fig:"               # Prefix assigned to figure label.
fig.path="figures/plot"     #
fig.scap="Short cap."       # Table of figures title.
fig.show="animate"          # Collect figures into an animation.
fig.show="hold"             #
fig.height=9                # Height of generated figure.
fig.width=12                # Width of generated figure.
include=FALSE               # Include code but not output/picture.
message=FALSE               # Do not display messages.
out.height=".6\\textheight" # Figure is 60\% of the page height.
out.width=".8\\textwidth"   # Figure is 80\% of the page width.
results="markup"            # Output from commands formatted.
results="hide"              # Do not show output from commands.
results="asis"              # Retain R output as \LaTeX{} code.
size="footnotesize"         # Useful for Beamer slides.
tidy=FALSE                 # Retain formatting used in the R code.
```

10.10 Exercises

Exercise 10.1

In the exercises for Chapter 7 we used a new self-identified sample dataset to report on data wrangling, predictive modelling, and a narrative from the data. Repeat the process but format the material within a LaTeX-based knitr document. RStudio will be a useful tool for this exercise.

Exercise 10.2

It was noted that R markdown provides a simplified approach to literate programming. Using the new and self-identified sample dataset from Chapter 7 replicate the data wrangling, predictive modelling, and narrative using R markdown in RStudio.

11

R with Style

Data scientists write programs to ingest, manage, wrangle, visualise, analyse and model data in many ways. It is an art to be able to communicate our explorations and understandings through a language, albeit a programming language. Of course our programs must be executable by computers but computers care little about our programs except that they be syntactically correct. Our focus should be on engaging others to read and understand the narratives we present through our programs.

In this chapter we present simple stylistic guidelines for programming in R that support the transparency of our programs. We should aim to write programs that clearly and effectively communicate the story of our data to others. Our programming style aims to ensure consistency and ease our understanding whilst of course also encouraging correct programs for execution by computer.

11.1 Why We Should Care

Programming is an art and a way to express ourselves. Often that expression is unique to us individually. Just as we can often ascertain who the author is of a play or the artist of a painting from their style we can often tell the programmer from the program coding structures and styles.

As we write programs we should keep in mind that something like 90% of a programmers' time (at least in business and government) is spent reading and modifying and extending other programmers' code. We need to facilitate the task—so that others can quickly come to a clear understanding of the narrative.

As data scientists we also practice this art of programming and indeed even more so to share the narrative of what we discover through our living and breathing of data. Writing our programs so that others understand why and how we analysed our data is crucial. Data science is so much more than simply building black box models—we should be seeking to expose and share the process and the knowledge that is discovered from the data.

Data scientists rarely begin a new project with an empty coding sheet. Regularly we take our own or other's code as a starting point and begin from that. We find code on Stack Overflow or elsewhere on the Internet and modify it to suit our needs. We collect templates from other data scientists and build from there, tuning the templates for our specific needs and datasets.

In being comfortable to share our code and narratives with others we often develop a style. Our style is personal to us as we innovate and express ourselves and we need consistency in how we do that. Often a style guide helps us as we journey through a new language and gives us a foundation for developing, over time, our own style.

A style guide is useful for sharing our tips and tricks for communicating clearly through our programs—our expression of how to solve a problem or actually how we model the world. We express this in the form of a language—a language that also happens to be executable by a computer. In this language we follow precisely specified syntax/grammar to develop sentences, paragraphs, and whole stories. Whilst there is infinite leeway in how we express ourselves and we each express ourselves differently, we share a common set of principles as our style guide.

The style guide here has evolved from over 30 years of programming and data experience. Nonetheless we note that style changes over time. Change can be motivated by changes in the technology itself and we should allow variation as we mature and learn and change our views.

Irrespective of whether the specific style suggestions here suit you or not, when coding do aim to communicate to other readers in the first instance. When we write programs we *write for others to easily read and to learn from and to build upon.*

11.2 Naming

1. Files containing R code use the uppercase `.R` extension. This aligns with the fact that the language is unambiguously called "R" and not "r."

   Preferred

   ```
   power_analysis.R
   ```

   Discouraged

   ```
   power_analysis.r
   ```

2. Some files may contain support functions that we have written to help us repeat tasks more easily. Name the file to match the name of the function defined within the file. For example, if the support function we've defined in the file is `myFancyPlot()` then name the file as below. This clearly differentiates support function filenames from analysis scripts and we have a ready record of the support functions we might have developed simply by listing the folder contents.

   Preferred

   ```
   myFancyPlot.R
   ```

   Discouraged

   ```
   utility_functions.R
   MyFancyPlot.R
   my_fancy_plot.R
   my.fancy.plot.R
   my_fancy_plot.r
   ```

3. R binary data filenames end in ".RData". This is descriptive of the file containing data for R and conforms to a capitalised naming scheme.

   Preferred

```
weather.RData
```

   Discouraged

```
weather.rdata
weather.Rdata
weather.rData
```

4. Standard file names use lowercase where there is a choice.

   Preferred

```
weather.csv
```

   Discouraged

```
weather.CSV
```

5. For multiple scripts associated with a project that have a processing order associated with them use a simple two digit number prefix scheme. Separating by 10's allows additional script files to be added into the sequence later.

   Suggested

```
00_setup.R
10_ingest.R
20_observe.R
30_process.R
40_meta.R
50_save.R
60_classification.R
62_rpart.R
64_randomForest.R
66_xgboost.R
68_h20.R
```

```
70_regression.R
72_lm.R
74_rpart.R
76_mxnet.R
80_evaluate.R
90_deploy.R
99_all.R
```

6. **Function names** begin lowercase with capitalised *verbs*. A common alternative is to use underscore to separate words but we use this specifically for variables.

   Preferred

   ```
   displayPlotAgain()
   ```

   Discouraged

   ```
   DisplayPlotAgain()
   displayplotagain()
   display.plot.again()
   display_plot_again()
   ```

7. **Variable names** use underscore separated *nouns*. A very common alternative is to use a period in place of the underscore. However, the period is often used to identify class hierarchies in R and the period has specific meanings in many database systems which presents an issue when importing from and exporting to databases.

   Preferred

   ```
   num_frames <- 10
   ```

   Discouraged

   ```
   num.frames <- 10
   numframes  <- 10
   numFrames  <- 10
   ```

8. **Function argument names** use period separated *nouns*. Function argument names do not risk being confused with class hierarchies and the style is useful in differentiating the argument name from the argument value. Within the body of the function it is also useful to be reminded of which variables are function arguments and which are local variables.

   Preferred

```
buildCyc(num.frames=10)
buildCyc(num.frames=num_frames)
```

   Discouraged

```
buildCyc(num_frames=10)
buildCyc(numframes=10)
buildCyc(numFrames=10)
```

9. Keep variable and function names shorter but self explanatory. A long variable or function name is problematic with layout and similar names are hard to tell apart. Single letter names like `x` and `y` are often used within functions and facilitate understanding, particularly for mathematically oriented functions but should otherwise be avoided.

   Preferred

```
# Perform addition.

addSquares <- function(x, y)
{
  return(x^2 + y^2)
}
```

   Discouraged

```
# Perform addition.

addSquares <- function(first_argument, second_argument)
{
  return(first_argument^2 + second_argument^2)
}
```

11.3 Comments

10. Use a single # to introduce ordinary comments and separate comments from code with a single empty line before and after the comment. Comments should be full sentences beginning with a capital and ending with a full stop.

    Preferred

```
# How many locations are represented in the dataset.

ds$location %>%
  unique() %>%
  length()

# Identify variables that have a single value.

ds[vars] %>%
  sapply(function(x) all(x == x[1L])) %>%
  which() %>%
  names() %T>%
  print() ->
constants
```

11. Sections might begin with all uppercase titles and subsections with initial capital titles. The last four dashes at the end of the comment are a section marker supported by RStudio. Other conventions are available for structuring a document and different environments support different conventions.

    Preferred

```
# DATA WRANGLING ----

# Normalise Variable Names ----

# Review the names of the dataset columns.

names(ds)

# Normalise variable names and confirm they are as expected.
```

```
names(ds) %<>% rattle::normVarNames() %T>% print()

# Specifically Wrangle weatherAUS ----

# Convert the character variable 'date' to a Date data type.

class(ds$date)
ds$date %<>%
  lubridate::ymd() %>%
  as.Date() %T>%
  {class(.); print()}
```

11.4 Layout

12. Keep lines to less then 80 characters for easier reading and fitting on a printed page.

13. Align curly braces so that an opening curly brace is on a line by itself. This is at odds with many style guides. My motivation is that the open and close curly braces belong to each other more so than the closing curly brace belonging to the keyword (`while` in the example). The extra white space helps to reduce code clutter. This style also makes it easier to comment out, for example, just the line containing the `while` and still have valid syntax. We tend not to need to foucs so much any more on reducing the number of lines in our code so we can now avoid Egyptian brackets..

    Preferred

```
while (blueSky())
{
  openTheWindows()
  doSomeResearch()
}
retireForTheDay()
```

Alternative

```
while (blueSky()) {
  openTheWindows()
  doSomeResearch()
}
retireForTheDay()
```

14. If a code block contains a single statement, then curly braces remain useful to emphasise the limit of the code block; however, some prefer to drop them.

Preferred

```
while (blueSky())
{
  doSomeResearch()
}
retireForTheDay()
```

Alternatives

```
while (blueSky())
  doSomeResearch()
retireForTheDay()
```

```
while (blueSky()) doSomeResearch()
retireForTheDay()
```

15. R is an interpretive language and encourages interactive development of code within the R console. Consider typing the following code into the R console.

```
if (TRUE)
{
  seed <- 42
}
else
{
  seed <- 666
}
```

After the first closing brace the interactive interpreter identifies a syntactically valid statement (an `if` with no `else`) and so executes it. The following `else` becomes a syntactic error. This will be true irrespective of whether we are interactively typing the commands directly into the R console or we are sending the commands from our editor in Emacs ESS or RStudio to the R console.

```
Error: unexpected 'else' in "else"

> source("examples.R")
Error in source("examples.R") : tmp.R:5:1: unexpected 'else'
4: }
5: else
   ^
```

This is not an issue when the if statement is embedded inside a block of code as within curly braces as we might use within a function definition. Here the text we enter is not parsed until we hit the final closing brace.

```
{
  if (TRUE)
  {
    seed <- 42
  }
  else
  {
    seed <- 666
  }
}
```

There is no simple solution for the interpreter so we might need to do something less satisfactory for top level statements in a script file or when writing interactively:

```
if (TRUE)
{
  seed <- 42
} else
{
```

```
  seed <- 666
}
```

16. Use a consistent indentation. I personally prefer 2 spaces within both Emacs ESS and RStudio with a good font (e.g., Courier font in RStudio works well but Courier 10picth is too compressed). Some argue that 2 spaces is not enough to show the structure when using smaller fonts. If it is an issue, then try 4 or choose a different font. We still often have limited lengths on lines on some forms of displays where we might want to share our code and about 80 characters seems about right. Indenting 8 characters is probably too much because it makes it difficult to read through the code with such large leaps for our eyes to follow to the right. Nonetheless, there are plenty of tools to reindent to a different level as we choose.

    Preferred

```
window_delete <- function(action, window)
{
  if (action %in% c("quit", "ask"))
  {
    ans <- TRUE
    msg <- "Terminate?"
    if (! dialog(msg))
      ans <- TRUE
    else
      if (action == "quit")
        quit(save="no")
    else
      ans <- FALSE
  }
  return(ans)
}
```

    Not Ideal

```
window_delete <- function(action, window)
{
          if (action %in% c("quit", "ask"))
          {
```

```
                        ans <- TRUE
                        msg <- "Terminate?"
                        if (! dialog(msg))
                                ans <- TRUE
                        else
                                if (action == "quit")
                                        quit(save="no")
                                else
                                        ans <- FALSE
            }
            return(ans)
}
```

17. Always use spaces rather than the invisible tab character.

18. Align the assignment operator for blocks of assignments. The rationale for this style suggestion is that it is easier for us to read the assignments in a tabular form than it is when it is jagged. This is akin to reading data in tables—such data is much easier to read when it is aligned. Space is used to enhance readability.

    Preferred

```
a       <- 42
another <- 666
b       <- mean(x)
brother <- sum(x)/length(x)
```

    Default

```
a <- 42
another <- 666
b <- mean(x)
brother <- sum(x)/length(x)
```

19. In the same vein we might think to align the `magrittr::%>%` operator in pipelines and the `base::+` operator for ggplot2 (Wickham and Chang, 2016) layers. This provides a visual symmetry and avoids the operators being lost amongst the text.

Such alignment though requires extra work and is not supported by editors. Also, there is a risk the operator too far to the right is overlooked on an inspection of the code.

Preferred

```
ds         <- weatherAUS
names(ds) <- rattle::normVarNames(names(ds))
ds %>%
  group_by(location) %>%
  mutate(rainfall=cumsum(risk_mm)) %>%
  ggplot(aes(date, rainfall)) +
  geom_line() +
  facet_wrap(~location) +
  theme(axis.text.x=element_text(angle=90))
```

Alternative

```
ds         <- weatherAUS
names(ds) <- rattle::normVarNames(names(ds))
ds                                                   %>%
  group_by(location)                                 %>%
  mutate(rainfall=cumsum(risk_mm))                   %>%
  ggplot(aes(date, rainfall))                          +
  geom_line()                                          +
  facet_wrap(~location)                                +
  theme(axis.text.x=element_text(angle=90))
```

20. An interesting variation on the alignment of pipelines including graphics layering is to indent the graphics layering and include it within a code block (surrounded by curly braces). This highlights the graphics layering as a different type of concept to the data pipeline and ensures the graphics layering stands out as a separate stanza to the pipeline narrative. Note that a period is then required in the `ggplot2::ggplot()` call to access the pipelined dataset. The pipeline can of course continue on from this expression block. Here we show it being piped into a `base::print()` to have the plot displayed and then saved into a variable for later processing.

Preferred

```
ds          <- weatherAUS
names(ds) <- rattle::normVarNames(names(ds))
ds %>%
  group_by(location) %>%
  mutate(rainfall=cumsum(risk_mm)) %>%
  {
    ggplot(., aes(date, rainfall)) +
      geom_line() +
      facet_wrap(~location) +
      theme(axis.text.x=element_text(angle=90))
  } %T>%
  print() ->
plot_cum_rainfall_by_location
```

11.5 Functions

21. Functions should be no longer than a screen or a page. Long functions generally suggest the opportunity to consider more modular design. Take the opportunity to split the larger function into smaller functions.

22. Generally prefer a single `base::return()` from a function. Understanding a function with multiple and nested returns can be difficult. Sometimes though, particularly for simple functions as in the alternative below, multiple returns work just fine.

    Preferred

```
factorial <- function(x)
{
  if (x==1)
  {
    result <- 1
  }
  else
  {
    result <- x * factorial(x-1)
```

```
  }

  return(result)
}
```

Alternative

```
factorial <- function(x)
{
  if (x==1)
  {
    return(1)
  }
  else
  {
    return(x * factorial(x-1))
  }
}
```

23. Align function arguments in a function definition one per line.
Preferred

```
showDialPlot <- function(label="UseR!",
                         value=78,
                         dial.radius=1,
                         label.cex=3,
                         label.color="black")
{
  ...
}
```

Discouraged

```
showDialPlot <- function(label="UseR!", value=78,
                         dial.radius=1, label.cex=3,
                         label.color="black")
{
  ...
}

showDialPlot <- function(label="UseR!",
```

```
value=78,
dial.radius=1,
label.cex=3,
label.color="black")
```

Alternative

```
showDialPlot <- function(
                      label="UseR!",
                      value=78,
                      dial.radius=1,
                      label.cex=3,
                      label.color="black"
                      )
```

24. Don't add spaces around the = for named arguments in parameter lists. Visually this ties the named arguments together and highlights this as a parameter list. This style is at odds with the default R printing style and is the only situation where I tightly couple a binary operator. In all other situations there should be a space around the operator.

    Preferred

    ```
    readr::read_csv(file="data.csv",
                    skip=1e5,
                    na=".",
                    progress=FALSE)
    ```

    Discouraged

    ```
    read.csv(file = "data.csv", skip =
             1e5, na = ".", progress
             = FALSE)
    ```

25. Arguments to function calls can also be aligned similarly to the function definition. An advantage of doing this is that all but the last argument can easily be commented out during testing of different options in using the function. An idiosyncratic alternative illustrated below places the comma at the beginning

of the line. This is actually particularly useful and works well to easily comment out specific arguments except for the first one. Often the first argument is the most important argument and is perhaps even a non-optional argument. Later arguments are often optional and we will explore different options and tune our code by commenting them in and out. This is quite a common style amongst SQL programmers and can be useful for R programming too.

Preferred

```
dialPlot(value=78,
         label="UseR!",
         dial.radius=1,
         label.cex=3,
         label.color="black")
```

Alternative

```
dialPlot(value=78
        , label="UseR!"
        , dial.radius=1
        , label.cex=3
        , label.color="black"
         )
```

Discouraged

```
dialPlot( value=78, label="UseR!",dial.radius=1,
         label.cex=3, label.color="black")
```

26. R has a mechanism (called namespaces) for identifying the names of functions and variables from specific packages. There is no rule that says a package provided by one author cannot use a function name already used by another package or by base R. Thus, functions from one package might overwrite the definition of a function with the same name from another package or from base R itself. A mechanism to ensure we are using the correct function is to prefix the function call with the name of the package providing the function, just like `dplyr::mutate()`.

Generally in commentary we will use this notation to clearly identify the package which provides the function. In our interactive R usage and in scripts we tend not to use the namespace notation. It can clutter the code and arguably reduce its readability even though there is the benefit of clearly identifying where the function comes from.

For common packages we tend not to use namespaces but for less well-known packages a namespace at least on first usage provides valuable information. Also, when a package provides a function that has the same name as a function in another namespace, it is useful to explicitly supply the namespace prefix.

Preferred

```
library(dplyr)      # Data wranlging, mutate().
library(lubridate) # Dates and time, ymd_hm().
library(ggplot2)    # Visualize data.

ds <- get(dsname) %>%
  mutate(timestamp=ymd_hm(paste(date, time))) %>%
  ggplot(aes(timestamp, measure)) +
  geom_line() +
  geom_smooth()
```

Alternative

```
ds <- get(dsname) %>%
  dplyr::mutate(timestamp=
                  lubridate::ymd_hm(paste(date, time))) %>%
  ggplot2::ggplot(ggplot2::aes(timestamp, measure)) +
  ggplot2::geom_line() +
  ggplot2::geom_smooth()
```

11.6 Assignment

27. Avoid using `base::=` for assignment. It was introduced in S-Plus in the late 1990s as a convenience but is ambiguous (named

arguments in functions, mathematical concept of equality). The traditional backward assignment operator `base::<-` implies a flow of data and for readability is explicit about the intention.

Preferred

```
a <- 42
b <- mean(x)
```

Discouraged

```
a = 42
b = mean(x)
```

28. The forward assignment `base::->` should generally be avoided. A single use case justifies it in pipelines where logically we do an assignment at the end of a long sequence of operations. As a side effect operator it is vitally important to highlight the assigned variable whenever possible and so out-denting the variable after the forward assignment to highlight it is recommended.

Preferred

```
ds[vars] %>%
  sapply(function(x) all(x == x[1L])) %>%
  which() %>%
  names() %T>%
  print() ->
constants
```

Traditional

```
constants <-
  ds[vars] %>%
  sapply(function(x) all(x == x[1L])) %>%
  which() %>%
  names() %T>%
  print()
```

Discouraged

```
ds[vars] %>%
  sapply(function(x) all(x == x[1L])) %>%
  which() %>%
  names() %T>%
  print() ->
  constants
```

11.7 Miscellaneous

29. Do not use the semicolon to terminate a statement unless it makes a lot of sense to have multiple statements on one line. Line breaks in R make the semicolon optional.

Preferred

```
threshold <- 0.7
maximum   <- 1.5
minimum   <- 0.1
```

Alternative

```
threshold <- 0.7; maximum <- 1.5; minimum <- 0.1
```

Discouraged

```
threshold <- 0.7;
maximum   <- 1.5;
minimum   <- 0.1;
```

30. Do not abbreviate `TRUE` and `FALSE` to `T` and `F`.

Preferred

```
is_windows  <- FALSE
open_source <- TRUE
```

Discouraged

```
is_windows  <- F
open_source <- T
```

31. Separate parameters in a function call with a comma followed by a space.

    Preferred

```
dialPlot(value=78, label="UseR!", dial.radius=1)
```

    Dicouraged

```
dialPlot(value=78,label="UseR!",dial.radius=1)
```

32. Ensure that files are under version control such as with github to allow recovery of old versions of the file and to support multiple people working on the same files.

11.8 Exercises

Exercise 11.1

Choose an R package from amongst those available on github (visit `https://github.com/cran` and choose a package). Perform a code walk-through and review for one or more of the R source files within the package. Produce a report with recommendations.

Exercise 11.2

Repeat this code walk-through for another package from github. Identify any differences in coding style and comment.

Exercise 11.3

There are a number of R coding style guides available on the Internet. Identify at least two others and compare them to the guide presented here. Critically review the differences and decide on which you personally might prefer. Write a report or blog post to compare and justify your choices.

Bibliography

Auguie B (2016). *gridExtra: Miscellaneous Functions for "Grid" Graphics.* R package version 2.2.1, URL https://CRAN.R-project.org/package=gridExtra.

Bache SM, Wickham H (2014). *magrittr: A Forward-Pipe Operator for R.* R package version 1.5, URL https://CRAN.R-project.org/package=magrittr.

Bates D, Maechler M (2017). *Matrix: Sparse and Dense Matrix Classes and Methods.* R package version 1.2-10, URL https://CRAN.R-project.org/package=Matrix.

BITRE (2014). "Ports: job generation in the context of regional development." *Information sheet*, Bureau of Infrastructure, Transport and Regional Economics, Department of Infrastructure and Regional Development. Material from the report used under the terms of the Creative Commons Attribution 3.0 Australia Licence available from http://creativecommons.org/licenses/by/3.0/au/deed.en, URL http://bitre.gov.au/publications/2014/files/is_056.pdf.

Breiman L, Cutler A, Liaw A, Wiener M (2015). *randomForest: Breiman and Cutler's Random Forests for Classification and Regression.* R package version 4.6-12, URL https://CRAN.R-project.org/package=randomForest.

Chamberlain S (2015). *ckanr: Client for the Comprehensive Knowledge Archive Network ('CKAN') 'API'.* R package version 0.1.0, URL https://CRAN.R-project.org/package=ckanr.

Chen T, He T, Benesty M, Khotilovich V, Tang Y (2017). *xgboost: Extreme Gradient Boosting.* R package version 0.6-4, URL https://CRAN.R-project.org/package=xgboost.

Dahl DB (2016). *xtable: Export Tables to LaTeX or HTML.* R package version 1.8-2, URL `https://CRAN.R-project.org/package=xtable`.

Durant W (1926). *The Story of Philosophy.* 2012 edition. Simon and Schuster.

Gagolewski M, Tartanus B, , other contributors; IBM, other contributors; Unicode, Inc (2017). *stringi: Character String Processing Facilities.* R package version 1.1.5, URL `https://CRAN.R-project.org/package=stringi`.

Grolemund G, Spinu V, Wickham H (2016). *lubridate: Make Dealing with Dates a Little Easier.* R package version 1.6.0, URL `https://CRAN.R-project.org/package=lubridate`.

Harrell Jr FE (2017). *Hmisc: Harrell Miscellaneous.* R package version 4.0-3, URL `https://CRAN.R-project.org/package=Hmisc`.

Henrion M (2007). "Open-Source Policy Modeling." *Journal of Law and Policy for the Information Society.*

Hocking TD (2017). *directlabels: Direct Labels for Multicolor Plots.* R package version 2017.03.31, URL `https://CRAN.R-project.org/package=directlabels`.

Knuth DE (1984). "Literate Programming." *The Computer Journal (British Computer Society)*, **27**(2), 97–111. URL `http://www.literateprogramming.com/knuthweb.pdf`.

Kuhn M (2017). *caret: Classification and Regression Training.* R package version 6.0-76, URL `https://CRAN.R-project.org/package=caret`.

Lobo-Pulo A (2016). "Evaluating Government Policies using Open Source Models." *Technical report*, Phoensight. URL `http://phoensight.com`.

Müller K, Wickham H (2017). *tibble: Simple Data Frames.* R package version 1.3.3, URL https://CRAN.R-project.org/package=tibble.

Neuwirth E (2014). *RColorBrewer: ColorBrewer Palettes.* R package version 1.1-2, URL https://CRAN.R-project.org/package=RColorBrewer.

R Core Team (2017). *R: A Language and Environment for Statistical Computing.* R Foundation for Statistical Computing, Vienna, Austria. URL https://www.R-project.org/.

Romanski P, Kotthoff L (2016). *FSelector: Selecting Attributes.* R package version 0.21, URL https://CRAN.R-project.org/package=FSelector.

Schloerke B, Crowley J, Cook D, Briatte F, Marbach M, Thoen E, Elberg A, Larmarange J (2016). *GGally: Extension to 'ggplot2'.* R package version 1.3.0, URL https://CRAN.R-project.org/package=GGally.

Sing T, Sander O, Beerenwinkel N, Lengauer T (2015). *ROCR: Visualizing the Performance of Scoring Classifiers.* R package version 1.0-7, URL https://CRAN.R-project.org/package=ROCR.

Soetaert K (2014). *diagram: Functions for visualising simple graphs (networks), plotting flow diagrams.* R package version 1.6.3, URL https://CRAN.R-project.org/package=diagram.

Therneau T, Atkinson B, Ripley B (2017). *rpart: Recursive Partitioning and Regression Trees.* R package version 4.1-11, URL https://CRAN.R-project.org/package=rpart.

Wickham H (2016). *scales: Scale Functions for Visualization.* R package version 0.4.1, URL https://CRAN.R-project.org/package=scales.

Wickham H (2017a). *stringr: Simple, Consistent Wrappers for Common String Operations.* R package version 1.2.0, URL https://CRAN.R-project.org/package=stringr.

Wickham H (2017b). *tidyr: Easily Tidy Data with 'spread()' and 'gather()' Functions.* R package version 0.6.1, URL https://CRAN.R-project.org/package=tidyr.

Wickham H, Bryan J (2017). *readxl: Read Excel Files.* R package version 1.0.0, URL https://CRAN.R-project.org/package=readxl.

Wickham H, Chang W (2016). *ggplot2: Create Elegant Data Visualisations Using the Grammar of Graphics.* R package version 2.2.1, URL https://CRAN.R-project.org/package=ggplot2.

Wickham H, Francois R, Henry L, Müller K (2017a). *dplyr: A Grammar of Data Manipulation.* R package version 0.7.0, URL https://CRAN.R-project.org/package=dplyr.

Wickham H, Hester J, Francois R (2017b). *readr: Read Rectangular Text Data.* R package version 1.1.0, URL https://CRAN.R-project.org/package=readr.

Williams GJ (1987). "Some Experiments in Decision Tree Induction." *Australian Computer Journal*, **19**(2), 84–91. URL http://togaware.com/papers/acj87_dtrees.pdf.

Williams GJ (1988). "Combining decision trees: Initial results from the MIL (multiple inductive learning) algorithm." In JS Gero, RB Stanton (eds.), *Artificial Intelligence Developments and Applications: Selected papers from the first Australian Joint Artificial Intelligence Conference, Sydney, Australia, 2-4 November, 1987*, pp. 273–289. Elsevier Science Publishers B.V. (North-Holland). ISBN 0444704655.

Williams GJ (2009). "Rattle: A Data Mining GUI for R." *The R Journal*, **1**(2), 45–55. URL http://journal.r-project.org/archive/2009-2/RJournal_2009-2_Williams.pdf.

Williams GJ (2011). *Data Mining with Rattle and R: The art of excavating data for knowledge discovery.* Use R! Springer, New York.

Williams GJ (2017). *rattle: Graphical User Interface for Data Mining in R.* R package version 5.0.14, URL `http://rattle.togaware.com/`.

Xie Y (2016). *knitr: A General-Purpose Package for Dynamic Report Generation in R.* R package version 1.15.1, URL `https://CRAN.R-project.org/package=knitr`.

Zhou ZH, Chawla NV, Jin Y, Williams GJ (2014). "Big Data Opportunities and Challenges: Discussions from Data Analytics Perspectives." *IEEE Computational Intelligence Magazine*, **9**(4), 62–74.

Index